Reactivity and Structure Concepts in Organic Chemistry

Volume 28

P. Heimbach · T. Bartik

An Ordering Concept on the Basis of Alternative Principles in Chemistry

Design of Chemicals and Chemical Reactions by Differentiation and Compensation

In cooperation with R. Boese, R. Budnik, H. Hey, A. I. Heimbach, W. Knott, H. G. Preis, H. Schenkluhn, G. Szczendzina, K. Tani and E. Zeppenfeld

With 122 Figures and 26 Tables

Springer-Verlag
Berlin Heidelberg New York
London Paris Tokyo Hong Kong

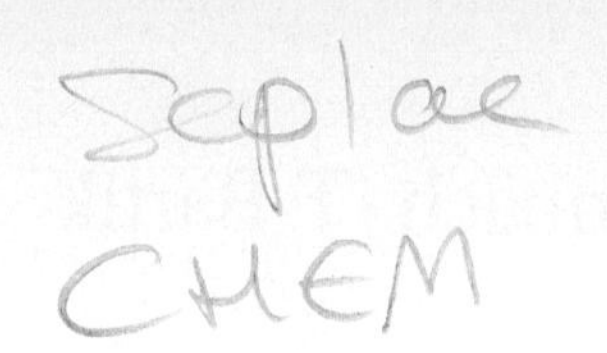

Professor Dr. Paul Heimbach
Dr. Tamás Bartik

Universität
Gesamthochschule Essen
Fachbereich 8 – Chemie
Postfach 10 37 64

4300 Essen 1

ISBN 3-540-51198-9 Springer-Verlag Berlin Heidelberg New York
ISBN 0-387-51198-9 Springer-Verlag New York Berlin Heidelberg

Library of Congress Cataloging-in-Publication Data

Heimbach, P. (Paul), 1934
An ordering concept on the basis of alternative principles in
chemistry: design of chemicals and chemical reactions
by differentiation and compensation / P. Heimbach, T. Bartik.
p. cm. – (Reactivity and structure: concepts in organic
chemistry; v. 28).
Includes bibliographical references.

1. Chemistry, Physical organic. I. Bartik, T. (Tamás), 1952.
II. Title. III. Series: Reactivity and structure; v. 28.
QD476.H35 1990 547.1′3–dc20 89-37149
ISBN 0-387-51198-9 (U.S.)

Printing: Mercedes-Druck, Berlin; Bookbinding: Lüderitz & Bauer, Berlin.
2151/3020-543210 – Printed on acid-free paper.

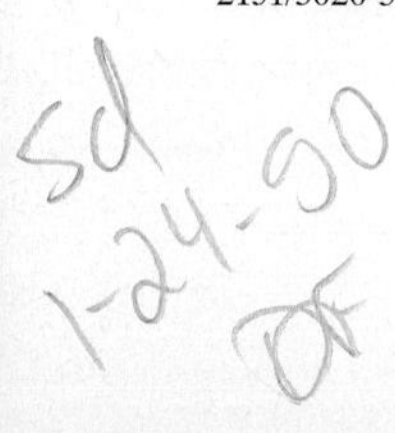

List of Editors

Professor Dr. Klaus Hafner
Institut für Organische Chemie der TH Darmstadt
Petersenstr. 15, D-6100 Darmstadt

Professor Dr. Jean-Marie Lehn
Institut de Chimie, Université de Strasbourg
1, rue Blaise Pascal, B.P. 296/R8, F-67008 Strasbourg-Cedex

Professor Dr. Charles W. Rees, F. R. S. Hofmann
Professor of Organic Chemistry, Department of Chemistry
Imperial College of Science and Technology
South Kensington, London SW7 2AY, England

Professor Dr. Paul v. Ragué Schleyer
Lehrstuhl für Organische Chemie der Universität Erlangen-Nürnberg
Henkestr. 42, D-8520 Erlangen

Professor Barry M. Trost
Department of Chemistry, The University of Wisconsin
1101 University Avenue, Madison, Wisconsin 53706, U.S.A.

Professor Dr. Rudolf Zahradník
Tschechoslowakische Akademie der Wissenschaften
J.-Heyrovský-Institut für Physikal. Chemie und Elektrochemie
Máchova 7, 12138 Praha 2, C.S.S.R.

Dedicated to all our teachers,
our coworkers
and especially to our families

Preface

Considering aspects of symmetry rules in chemistry, one is faced with contradictory terms as for example, "90% concertedness" sometimes being used in literature. To accept conservation of orbital symmetry to be as controlled as inversion by alternative principles seems far more promising. The intention of this book is aimed at introducing a qualitative understanding of phase relations in electromagnetic interactions. Avoiding one-sided dogmatism we tried to demonstrate the importance of alternative principles as guidelines to the evolution of alternative order in chemical systems.

Passing through the jungle of information it became extremly important to control again and again our insights into the ordering phenomena by experiments under conditions as coherent as possible. We became more aware of the fact that chemistry – the science of "becoming" in complex systems – can not be understood by mechanistic details, i.e. THROUGHPUT-studies alone, because the mechanism is only true for the special system under investigation and does not offer a tool for the evolution of opposite order.

We had to accept chemistry as a mediator between molecular physics and general epistemology. This quite unusual combination was directed by excellent teachers and the realizations were made possible by enthusiastic, open-minded coworkers (see references). The next target we will strive for on this journey will be to quantify the alternative principles, that means obtaining the order parameters of H. Haken (e.g. in asymmetric synthesis).

For this task I hope to meet more new young friends as I found in the past with H. Schenkluhn and T. Bartik. My co-author returned to Hungary too soon and therefore I have to write this preface alone.

Nevertheless this book is a consequence of excellent team work of both my research groups at the University of Essen and as guests at the Max-Planck-Institut für Kohlenforschung in Mülheim. For generous hospitality since 1973 my coworkers and I cordially thank Prof. Dr. G. Wilke, the director of the institute in Mülheim.

The original title of the book “THE pragmatic CONCEPT OF alternative and alternating ORDER FACTORS” and the subtitle were changed following proposals of Yoshiharu Izumi, Professor emeritus of the OSAKA UNIVERSITY and one of the authors of “Stereodifferentiating Reactions, the nature of asymmetric reactions”, Academic Press. Fundamental discussions were made possible at international conferences in KOBE and KYOTO sponsored by the TANIGUCHI FOUNDATION.

My secretary Mrs. I. Reiter was a permanent and skilful helper.

Essen, September 1989 P. Heimbach

Contents

Glossary XV

1 Introduction 1

2 Characterization of Substituents by Patterns and Recognition of ALTERNATIVE PRINCIPLES 9

2.1 Correlation with the Parameter Sets Θ and χ by Tolman 13
2.2 Systematic Variations at Tetrahedral Centers: A New Parameter Set Δ 14
2.3 The Importance of the Representative Substituents -OMe/-SMe and -CMe_3/-$SiMe_3$ 18
2.4 The Importance of P- and Δ-parameter Sets for Pattern Comparison 24
2.5 The Importance of Compensation Phenomena 24
2.6 Further Characterization of Substituents by Pattern Comparison 28
2.7 Conformational Changes in Phenyl Systems 32

3 Examples of Absolute, Alternative Orders in Chemical Systems by Pairs and Alternating Classes of ALTERNATIVE PRINCIPLES 37

3.1 Examples for the Absolute, Alternative Effects by ACC/DO Heteroatoms 37
3.2 Separation of Main Group and Transition Metal Elements in Four Sectors (PSE-sectors) 40
3.3 Errors of Logical Typing in DO/ACC Alternatives 47
3.4 The Alternative Principles EVEN/ODD 47
3.5 Aspects of Coupling Chemical Subsystems: The Alternative Principles OPEN/CYCLIC 52
3.6 The Metala-Logy Principle 55
3.7 The Principle of Alternative Positions 60
3.8 The Alternative Principles IONIC/COVALENT 61

4 Representation of Differentiation and Compensation of ALTERNATIVE PRINCIPLES 67

4.1 On the Definition of Paritetic and Complementary ALTERNATIVE PRINCIPLES and their Effects 67
4.2 Differentiation and Compensation of Two Pairs of ALTERNATIVE PRINCIPLES 70
4.3 Alternative Patterns by Classes of ALTERNATIVE PRINCIPLES . . 73
4.4 Symmetric/Antisymmetric Coupling of Two Pairs of ALTERNATIVE PRINCIPLES 76
4.5 Representation in Hierarchically Ordered, Multi-dual Decision-trees 80

5 Representative Examples of Multi-dual Decision – Trees: A Generalization of Phase Relation Rules 83

5.1 The Control of Elimination Reactions by ALTERNATIVE PRINCIPLES 83
5.2 The Alternative Control of Knoevenagel Versus Michael Reaction of Mesityloxide 87
5.3 Control of Asymmetric Synthesis in a Metal-induced Ketone Synthesis 92
5.4 Influence on Structures and Processes in Transition Metal Complexes 103
5.5 A Comparison of the Catalytic Oligomerizations of Propanal 107

6 The Discontinuous Method of INVERSE TITRATION 109

6.1 Evidence in Support of Concentration Effects: A Summary 109
6.2 Examples for the Application of INVERSE TITRATION 117

6.2.1 Application of INVERSE TITRATION in Metal Catalysis 117
6.2.2 Application of the INVERSE TITRATION in Organic Syntheses 122
6.2.3 Application of INVERSE TITRATION in SYSTEM ENLARGEMENT 127
6.2.4 Application of INVERSE TITRATION for the Proof of the Influence of Catalyst Poisons 130
6.2.5 Aspects in Application of INVERSE TITRATION for the Ni-ligand Modified Catalytic Propene Dimerization 132

6.3 Outlook and Unsettled Problems 134

7 Molecular Architecture: Some Definitions 136

7.1 Structure of the Whole System 136
7.2 Intermolecular SYSTEM ENLARGEMENT 138
7.3 Intramolecular SYSTEM ENLARGEMENT and VARIATION: Substitution of Hydrogen by Substituents and Carbon by Heteroatoms 142
7.4 Coupling of Subsystems 144
7.5 Symmetry Aspects in the Coupling of Chemical Subsystems 147

8 Models and Methods for the Understanding of Self-organization and Synergetics in Chemical Systems 149

8.1 The Ordering CONCEPT OF ALTERNATIVE PRINCIPLES in a Comprehensive Form 149
8.2 Some Statements to the Application of the CONCEPT OF ALTERNATIVE PRINCIPLES 153
8.3 Application of the CONCEPT OF ALTERNATIVE PRINCIPLES as Compensation Strategy 156

9 Information from Alternatives in Biochemistry 165

9.1 Alternative Information on Nucleic Acids and α-Amino-Carboxylic Acids 165
9.2 Alternative Information at Bio-membranes 169
9.3 Chirality, an Error in Logical Typing 172
9.4 Restriction of the Number of Realizations in Evolved Systems 172

10 Acknowledgements and Petition 175

11 Appendix 181

12 References 186

13 Epilogue 195

Subject Index 211

Glossary

Alternating phenomena are observed in continuous series of alternative principles, when EVEN/ODD numbers e.g. of valence electrons as in C/N/O/F, of π-electron pairs as in ethene/buta-1,3-diene/hexa-1,3,5-triene/octa-1,3,5,7-tetraene or of main quantum numbers as in F/Cl/Br/I play a significant role depending on the system.

Alternative principles (order factors) lead to an inverse absolute or relative behavior in chemicals and chemical processes. We separate them into classes of paritetic and complementary principles. Nevertheless both types are considered to contribute to either positive or negative (opposite) differentiations (cooperations) or more or less similar compensations caused by phase relation rules in electromagnetic interactions. The alternative principles are represented in this monograph by black and white or generally by stars (☆/★ or ★/☆) or in unified symbols (e.g. ★ or ★). Alternatives in the widest sense are separated by one dash (/) two alternatives by two dashes (//). Alternative pathways or decisions are marked by black and white arrows.

Compensation phenomena may destroy information of alternative principles by – often unexpected – self-organization. But applied as a strategy, compensation phenomena may help to reactivate a system perturbated by an alternative principle by inverse ones or to circumvent patented chemicals and chemical processes by corresponding alternatives in subsystems.

Complementary alternatives. These alternatives are not describable by paritetic principles alone. In addition one has to consider for example changes in energy content (changing for example the standard of the alternatives).

In extreme cases we even mix alternative principles out of totally different concepts like alternatives in molecular arrangements and stochastic models like MUCH/A LITTLE in thermodynamics or localized charge (+/−) and S/AS-coupling in electromagnetic field interactions so far without running into problems of errors in logical typing.

Differentiation in opposite directions or to opposite orders is effected by two (or four) combinations of two pairs of alternative principles in subsystems.

NAMES AND TERMS USED IN CHEMISTRY

EITHER	OR
DETERMINED BY DEFINITION AND HISTORY	DETERMINED BY MEANING AND UNDERSTANDING
CONSTANTS FOR SUBSTITUENTS, EQUILIBRIA etc. (only algorism constant)	PARAMETERS FOR SUBSTITUENTS, EQUILIBRIA etc. (property dependent)
ENTROPY/NEGENTROPY (used by Schrödinger, Bateson ...)	+/- ENTROPY (determined by sign)
"DO/ACC"-SUBSTITUENTS (too less differentiated)	SUBSTITUENTS (classified by alternatives; DO/ACC, EVEN/ODD, 4q-/4q+2- etc.)
ALLOWED/FORBIDDEN (Phase relation rules, orbital symmetry rules in absolute terms)	ACTIVATED/INHIBITED and two times compensated (Phase relation rules in relative terms)
CHIRALITY left/right (but only one parity in both hands)	PARITY $P=+1_{ref.}/-1$ (mirror image in odd dimensions)
KEY/LOCK HANDS/GLOVES (only space)	DIASTEREOMERIC INTERACTION (+1 +1) / (+1 -1) / (-1 +1) / (-1 -1) (space and time)
electronic/steric (electromagnetic) (mechanic) (microscopic) (macroscopic)	attractive/repulsive (both electromagnetic)
DOUBLE STEREODIFFERENTIATION (only for R/S combinations)	ORDERING CONCEPT ON THE BASIS OF ALTERNATIVE PRINCIPLES (for all alternatives, when parity is broken)
ALLOSTERIC (alternatives in space)	ALLOPHASIC (alternatives in phase relations)
INTER-/INTRA-MOLECULAR (only molecular level)	INTER-/INTRA-MODAL (on the level of HAKEN'S modes)

The type of symmetric or antisymmetric coupling of the two combinations may be recognized by pattern comparison. The remaining two other combinations always lead to more or less expressed compensations phenomena.

Errors of logical typing occur when definitions and names are used for members in classes of different logical types e.g. DO/ACC for individual atoms or for ensembles of atoms – substituents – like $-OCH_3$/$-COOCH_3$.

Evolution. This term is used to designate the process of developing out of alternative principles the pathway to only one harmonic unity, like a pure chemical or an individual chemical reaction. The term is not used to show that chemical species or reactions are derived from each other.

An ordering concept of alternative principles. This concept is developed to get reductionistic prescriptions for the next alternatives in the design of a holistic unity. It is useful for experimenters and theoreticians.

Paritetic alternatives are e.g. absolute alternatives in charge (+/−), local arrangements (R/S) and phase relations (S/AS) without any further differences.

Patterns combine different quantities (analogous data) to an alternative (digital) "Gestalt" by allowing the recognition of hidden qualities. Pattern recognition is an important tool for elucidating the order determining qualities in complex systems.

1 Introduction

Order in the microworld produces strength in the macroworld.

H. Haken [1]

The experience gained during experimental investigations in the course of years to find out the scope of metal-induced and – catalyzed organic synthesis has taught us a procedure in the investigation of complex chemical systems.

In spite of defining our

	S	ystems	(objects of investigations)
	M	odels	(of thinking)
and	M	ethods	(of experimental procedures)

very extensively and exactly, we always needed to introduce further variations displaying normally inverse (i.e. alternative) character in the system or subsystems under investigation

by

changing	T	ype[1]	(of inverse perturbations in "pairs")
	A	mount	(of subsystems)
	P	osition	(bearing relevance in the special system)
and	S	patial arrangement	(of inverse parity).

Only *by experiments* did we recognize the relations between subsystems arising from alternatives by differentiations and compensations in the whole behavior and we were finally able to transfer our experience by means of analogy to further chemical systems. In order to get a definitive insight into the behavior of chemical systems – being always much more complex than generally assumed – by systematic and experimentally realizable variations it was of significant importance to make use of information and the mental strategy in related disciplines [2,3,4] or interdisciplinary sciences [5,6,7]. These ideas are represented in popular written books of natural philosophy or epistemology [8,9] (vide Chap. 10).

To demonstrate the target of this monograph by a few problems we have selected four examples in Fig. 1.1. In Fig. 1.1a we point to a certain contradiction caused by the fact that the methoxy group as substituent of a double bond makes

[1] Burning new problems demand, at the beginning, experimentalists at the *taps*, afterwards normally theoreticians are the tops.

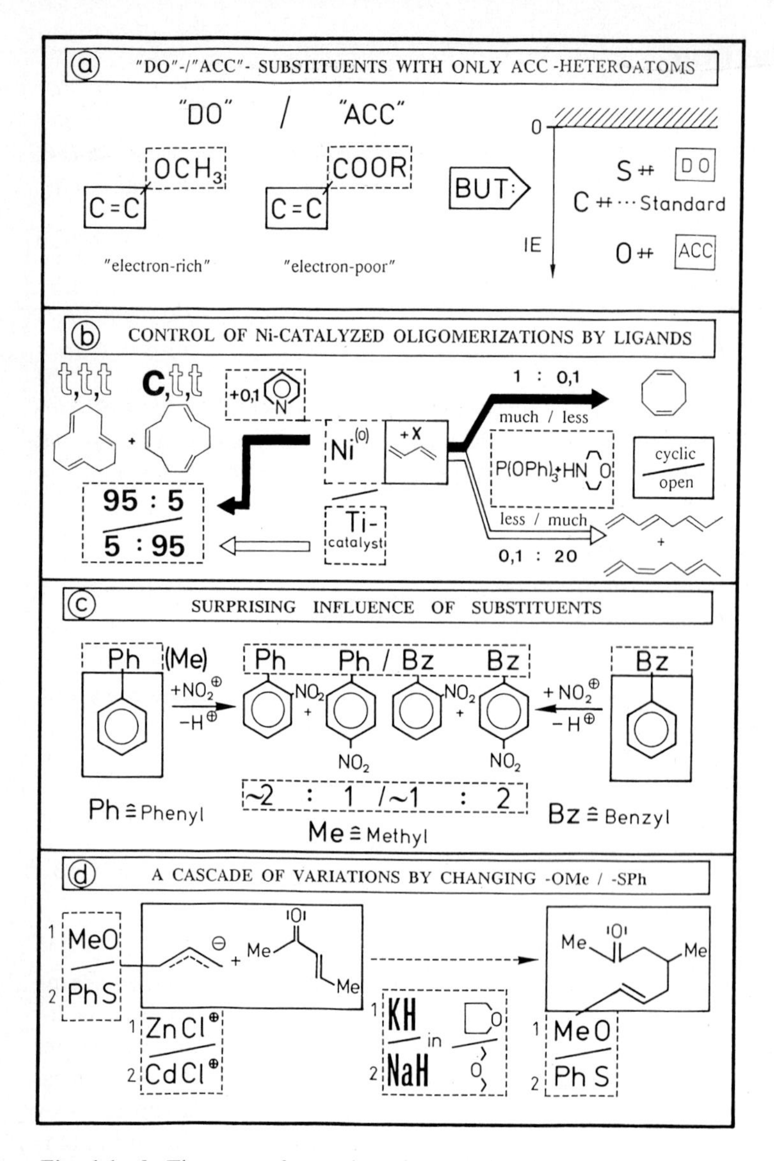

Fig. 1.1a-d. There are four selected examples, **(a)-(c)** demonstrate what alternative decisions in subsystems (marked by *broken lines*) provoke alternatives in reactivity, reaction rates (activation/inhibition) or in selectivities of the products (for definition of selectivity see Fig. 8.1). The last example in **(d)** demonstrates the alternatives in reaction cascades (see text)

it electron rich and the carboxylic acid ester group shows the inverse effect leading to an electron deficient π-system. This is observed only with oxygen as heteroatom, which, compared with carbon, is always the better acceptor of electrons. In Chap. 2 we will therefore discuss the problems of how to characterize substituents more intensively and show the way to avoid errors in logical typing because the terms for individuals (atoms: DO/ACC) and classes (substituents as classes of atoms: "DO/ACC") have to be chosen separately. *Terms are always set in parentheses, when errors in logical typing may arise by false classification.* The examples in (b) demonstrate the importance of choosing the catalyst metal with respect to the stereoselectivity of the otherwise identical product (Chap. 3) [10–12]. At the same time the importance of the chosen amount can be illustrated – namely MUCH/A LITTLE or alternatively A LITTLE/MUCH – of two identical effectors (triphenylphosphite and morpholine, vide Chap. 6) to convert the substrate butadiene at a $Ni^{(o)}$-catalyst either to an eight-membered ring (COD) or to an openchained dimer (*n*–OT) [13].

Surprising inverse effects in the nitration of aromatic compounds are caused by the choice of phenyl or benzyl substitution (vide Fig. 1.1c). The statistically expected *ortho/para* ratio of $\approx$ 2:1 in the case of diphenyl is accompanied by an inversion of this regioselectivity in the nitration of diphenylmethane [14] (vide Chap. 2).

For example (d) in Fig. 1.1 it is pointed out that by only one change in a subsystem (of the substituent $-OCH_3$ by $-SC_6H_5$) a whole cascade of variations in a multistep synthesis is necessary [15] to compensate for this single perturbation (for this vide Chaps. 3, 5 and 8). All legends of the figures are considerably shortened so that full information about the meaning in the figures of high abstraction is only available by reading the accompanying text – and step by step through the increasing knowledge of the reader. For formal representation of the results in the surely unfamiliar form of patterns see Chap. 4.

The authors can not guarantee a thorough working up of all the information in the literature concerning all aspects according to the presented CONCEPT OF ALTERNATIVE PRINCIPLES. As alternative principles we designate all factors which lead to an inverse order. All these factors must be chosen in reality by the experimenter (vide Chap. 8). The term order is not related here to formal kinetics but to alternatives with respect to space, time, symmetry (for this term see [16] p 124 and [17]) and so on. The citation of the whole literature is made even more difficult by the fact that very often only the "best" and not all experimental information is published in detail and that, at the same time, frequently more than one alternative principle in the chemical system or subsystem under investigation is varied at the same time without demonstrating their cooperativities and compensations.

The reason and justification for taking the examples resulting from our experimental work and the experiments of our coworkers in order to demonstrate general phase relation is not based on the fact that we overestimate these contributions but, that we are familiar with those systems and their conditions of experimental coherency investigated by us or our staff applying this ordering

CONCEPT OF ALTERNATIVE and alternating PRINCIPLES [18–29]. In Fig. 1.2, in each case two chemical systems belonging to organic and metalorganic chemistry are selected in which we extensively applied the preparative strategies to optimize and to control different selectivities as presented in this monograph. In example (a), is assigned by broken lines, the subsystems in which the alternative variations were carried out to control for example the synthesis of *cis-* or *trans*-olefines (i.e. the stereoselectivity) in the Wittig reaction [30–32]. The very intensive efforts of many of our colleagues can not, unfortunately, be taken into consideration because a literature search on this subject performed by courtesy of Bayer AG, found about 500 publications and that was only for 1980 to 1981. As examples we present only [33,34]. In (b), the control of *ortho-/para*-substitution of monosubstituted benzene-derivatives is discussed for the nitration without mentioning the – here unimportant – *meta*-substitution [14,35]. Referring to the system presented in (c), the synthesis of the two enantiomers of the β,γ-unsaturated 3-methyl-E-4-hexen-2-one will be discussed in Chap. 5. Thereby it was our aim to control the direction and extension of the optical induction by introducing subsystems of only one parity (derived from the "chiral pool" [36,37] "of nature") and by changing so called "achiral alternative principles". Whenever we examine a system under coherent conditions we were aware of the two components system in (d) Ni/butadiene exerting control of eight/six-membered ring synthesis by means of the amount and type of added P-ligands. This system has offered an instructive tool over the years [18–29,38–40].

Interesting and informative books and publications on perturbation-theory [41], HMO-theory [42], FMO-theory [43,44] and aspects of molecular architecture [16,45] and titration processes [46,47] in chemistry lead to a deeper understanding of the general effects of additional perturbations introduced into a given chemical system. This information is a good and necessary – but here not discussed – prerequisite for elucidating and understanding the background and basis of the additional experimental strategies presented in this book (vide Chap. 8).

The aim of this monograph is to make available to the synthetic chemist some alternative facts of theoretical chemistry which have, in principal, been well known for a long time as experimental strategies and concepts of organic synthesis (vide Chap. 8). The CONCEPT OF ALTERNATIVE PRINCIPLES [23,29] – as an integrating idea to avoid ideology (one-sided aspect) – is in part well known, too, and should not replace the other older but useful concepts of chemistry but will merely supply those models and look "at both sides of the same medal".

The four examples in Fig. 1.3 concerning six-membered ring syntheses are of general interest and meet the demands of illustrating the disadvantages arising from one-sided considerations both of ionic interactions, as with the acid catalyzed six-membered ring formation, or the Robinson anellation and of covalent interactions in ring closure of π-systems [48] or in the Diels-Alder reaction [49] for many aspects of these syntheses. All interactions are, indeed, always important. By neglecting one of the two aspects phenomena will arise which cannot be understood. Some of these are introduced (vide column 3 of Fig. 1.3) in this

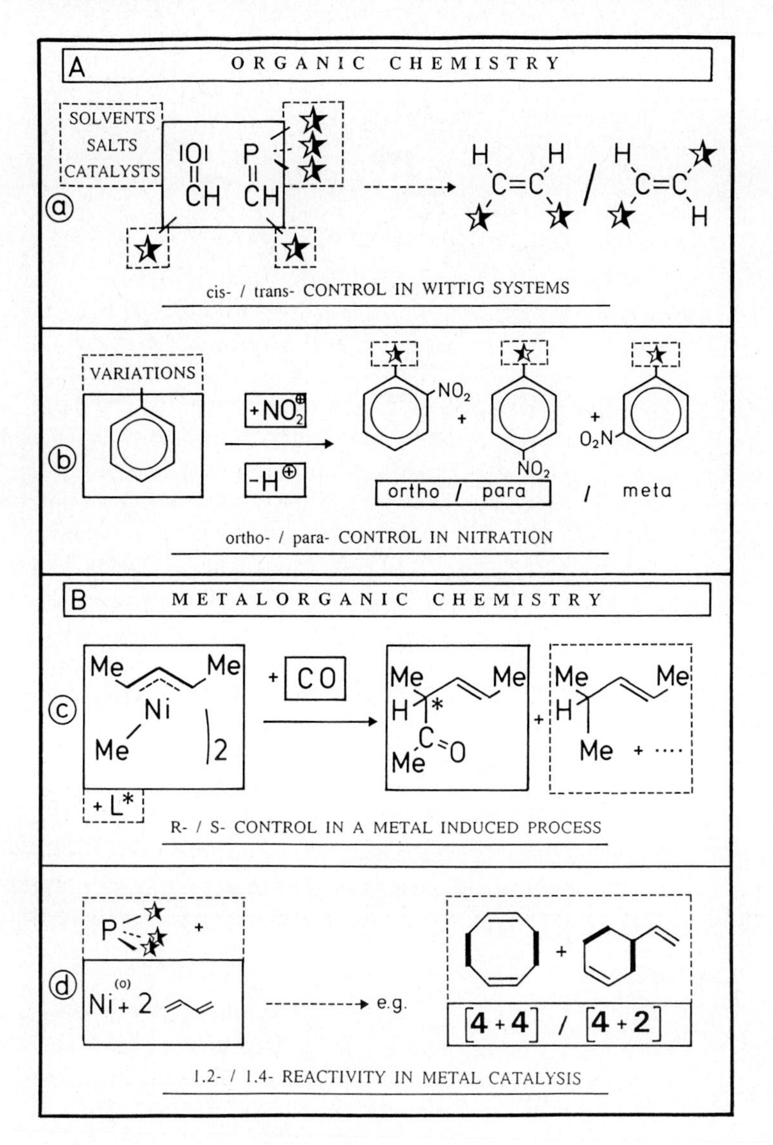

Fig. 1.2a-d. Alternatively ordering factors, e.g. marked *black/white* (vide Chap. 3), in subsystems (surrounded by *broken lines*) may control either stereoregio-, enantio- or chemo-selectivity in the Wittig reaction **(a)** [30–32], the nitration of substituted arenes **(b)** [35], the metal-induced asymmetric synthesis **(c)** and in the ligand modified metal catalysis **(d)**

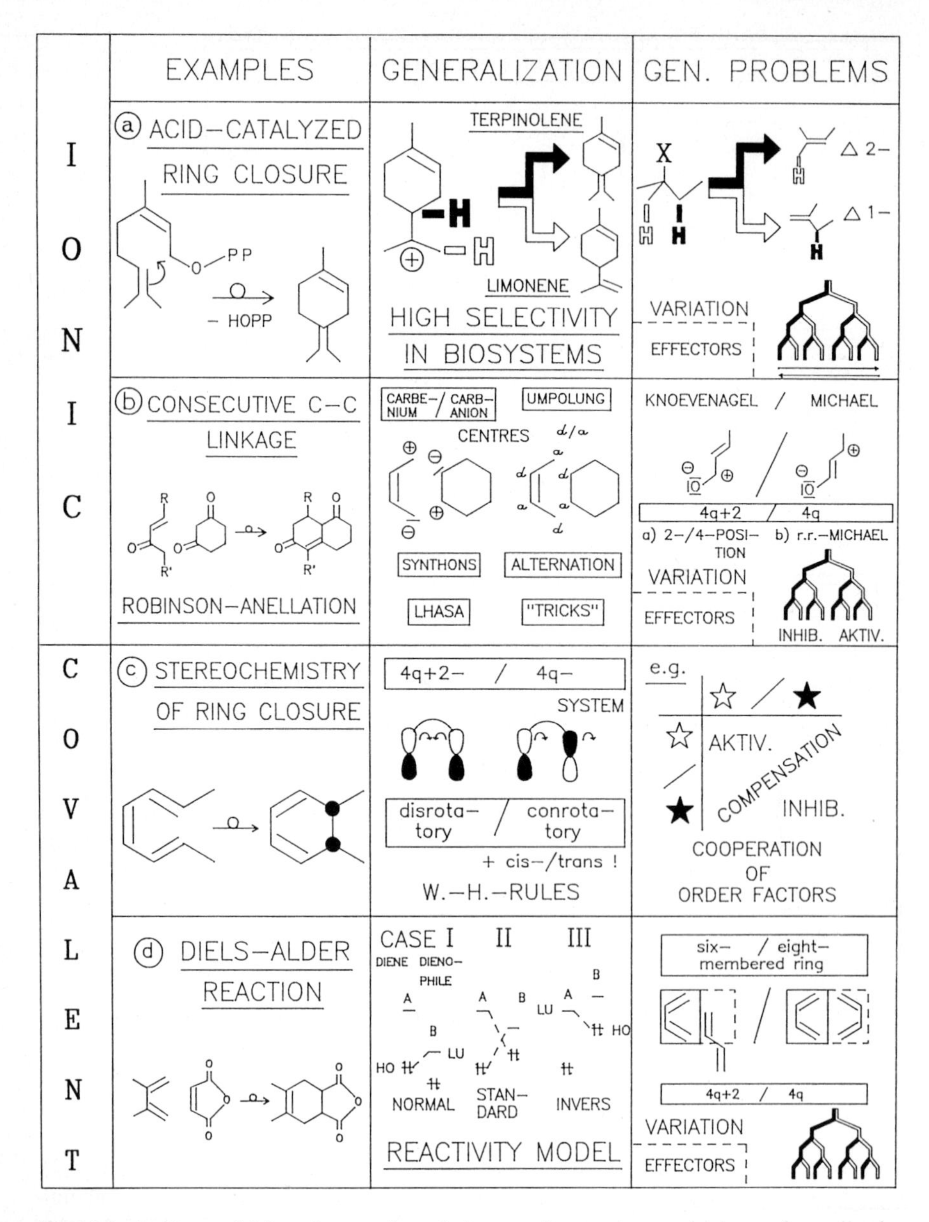

Fig. 1.3a-d. Examplifying six-membered ring syntheses alone, which are formally designated as charge, **(a)** and **(b)**, or orbital controlled, **(c)** and **(d)**, reactions, generalizations will be drawn (column 2); possible problems in partial aspects are demonstrated in column 3 and will be discussed later in detail step by step

monograph in the form of questions of general interest in subsystems and alternatives which will be discussed in detail later on.

The choice of the investigated systems – as seen in the four examples – has been selected by our teams for some years according to the following principles [50]:

1. The reaction to be investigated should be important, either in undergraduate education in organic chemistry or in industry.
2. Simultaneously, it should be related to fundamental problems in biochemical systems.

Additional aspects for selection with respect to this monograph are as follows:

3. The reactions should elucidate dubious definitions and point out preconceptions and errors in logical typing.
4. The CONCEPT OF ALTERNATIVE PRINCIPLES should be applied intensively in the elucidation of the corresponding system.

By applying these criteria, we have shown in this introduction that pragmatic models and experimental methods for the evolution of even complex chemical systems are possible. Biochemical systems are introduced, in addition, to enlarge the deep experience of experimentalists. On the other hand, the experimental results of relatively simple systems should imply a comprehension of the perfect control in biochemical systems (Chap. 9 in cooperation with A. Heimbach). We are experimentalists, who would like to express our thanks to theorists with this monograph (Chap. 10).

In Fig. 1.4, how alternatives in chemistry with increasing complexity may be controlled by ordering inverse principles is demonstrated. Experimental results as a consequence of alternative variations of ordering principles in subsystems may supply the answer to the problem of the origin of "all forms" as a reduction to electromagnetic interactions, provided by combinations of irreducible simple alternatives [3].

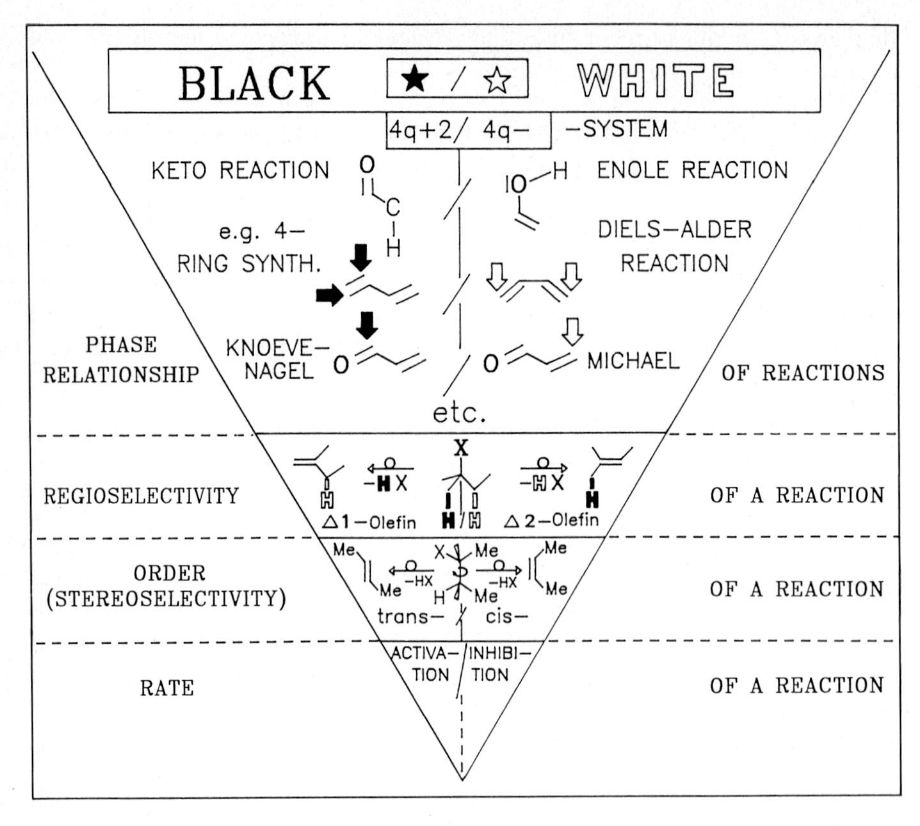

Fig. 1.4. Starting from activation/inhibition phenomena of a given reaction, passing alternative selectivities of a reaction and reaching general symmetry control in alternative reactivity of a 4q–/4q+2-system, it will be demonstrated, that systems of increasing complexity may be controlled by symmetry adapted cooperation of pairwise alternatives (vide Chap. 3 and 5): experimental evolution. For a formal description see Chap. 4

2 Characterization of Substituents by Patterns and Recognition of ALTERNATIVE PRINCIPLES

The sort of thinking that deals with quantity resembles in many ways the thinking that surrounds the concept of energy; whereas the concept of number is much more closely related to the concepts of pattern and negentropy

G. Bateson [9]

Up to 1979 about 5.000–10.000 applications of FE- (*f*ree *e*nergy) or LFE- (*l*inear *f*ree *e*nergy) relations in organic chemistry are described in the literature. This distinctly demonstrates the importance of detection and description of quantitative changes e.g. in chemical reactivity of parent systems (a determining and invariant constituent of a whole system). Hammett-, Taft- and other parameters [51,52] are applied to account for the effects of variable substituents in carbon-σ- and -π-systems. More seldom, corresponding parameters are put to the successful description of quantitative changes in kinetic or thermodynamic values of metal-induced or -catalyzed processes (see e.g. [53–55]). Within the framework of this chapter, we will, at the beginning, restrict ourselves to the application of those parameters referring to the properties of phosphanes and phosphites (P-ligands). Overwhelmingly, we will demonstrate, by pattern recognition, that we are able to characterize in more detail the ordering power of organic substituents.

C.A. Tolman [56] risked a new start in metalorganic chemistry and proposed a "steric" and "electronic" parameter for P-ligands, which control the behavior of metal-induced and -catalyzed processes (vide Chap. 6) by associations with the metal. For definitions see Scheme 2.1.

In the range of the experimental error of his method (+/−1 cm^{-1}) Tolman found out, that in P-ligands the influences of their substituents are additive with respect to the χ- and Θ-parameters. Therefore incremental Θ_i- and χ_i- values can be calculated for individual substituents to characterize the "electronic" and "steric" contributions (for this see Fig. 3.23) of these substituents at the phosphorus.

$$\Theta_{i(X)} = \Theta_{PX_3}/3 \qquad \chi_{i(X)} = \chi_{PX_3}/3$$

Therefore is e.g. $\chi_{PX^1X^2X^3} = \chi_{1(X^1)} + \chi_{2(X^2)} + \chi_{3(X^3)}$

We would like to demonstrate here some applications of these parameters for substituents. The "steric" parameters Θ_i and the "electronic" parameters χ_i are proportional to energy values according to their definitions as it is true for the corresponding parameters of organic chemistry e.g. σ_I [57] and E_s^N [58]. As pointed out elsewhere [20,59] and shown in Fig. 2.1 by examples, the Θ_i- and χ_i-values of the substituents correlate very well with their "steric" respectively "electronic" parameters applied in organic chemistry (E_s^N and σ_I).

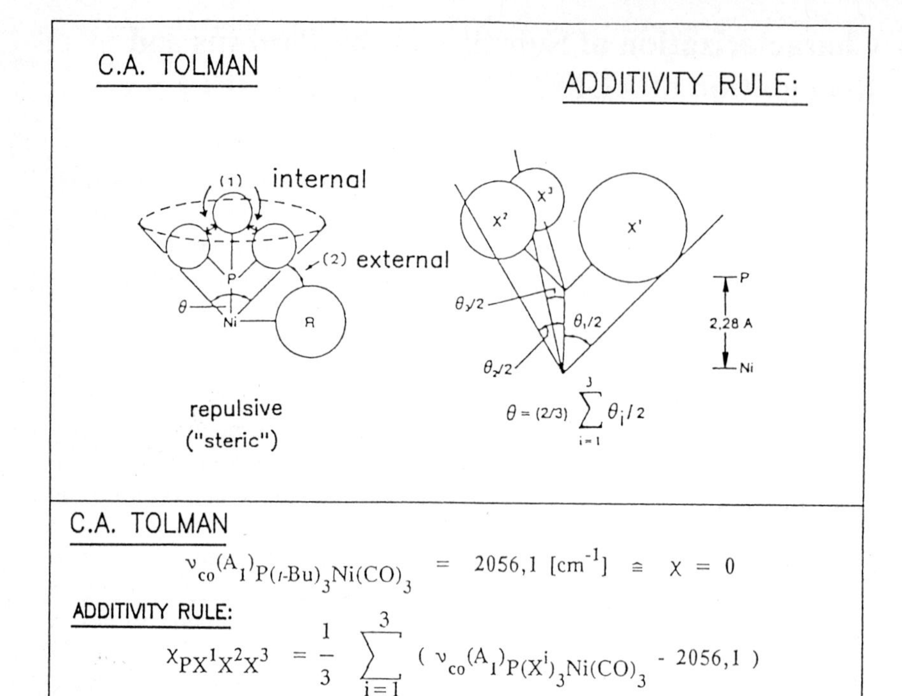

T. Bartik

$$\nu_{co}(A_1)Ph_3PNi(CO)_3 = 2069{,}35\ [cm^{-1}] \mathrel{\hat{=}} P_L = 0$$

Scheme 2.1

Conclusion: The Θ_i- and χ_i-values may be quite generally considered to be parameters for substituents and may be used instead of σ_I and E_s respectively, the values being convertible.

The "steric" parameter Θ was defined as space filling by phosphorus ligands caused by the substituents at the phosphorus, here as cone angle of models with the respective substituents.

The top of the cone terminates in the metal atom. The "electronic" parameter is defined as the difference in wave numbers of the symmetric (A_1)-carbonyl stretching frequency of L-$Ni(CO)_3$ complexes of the chosen P-ligand (υ_{co} (A_1))

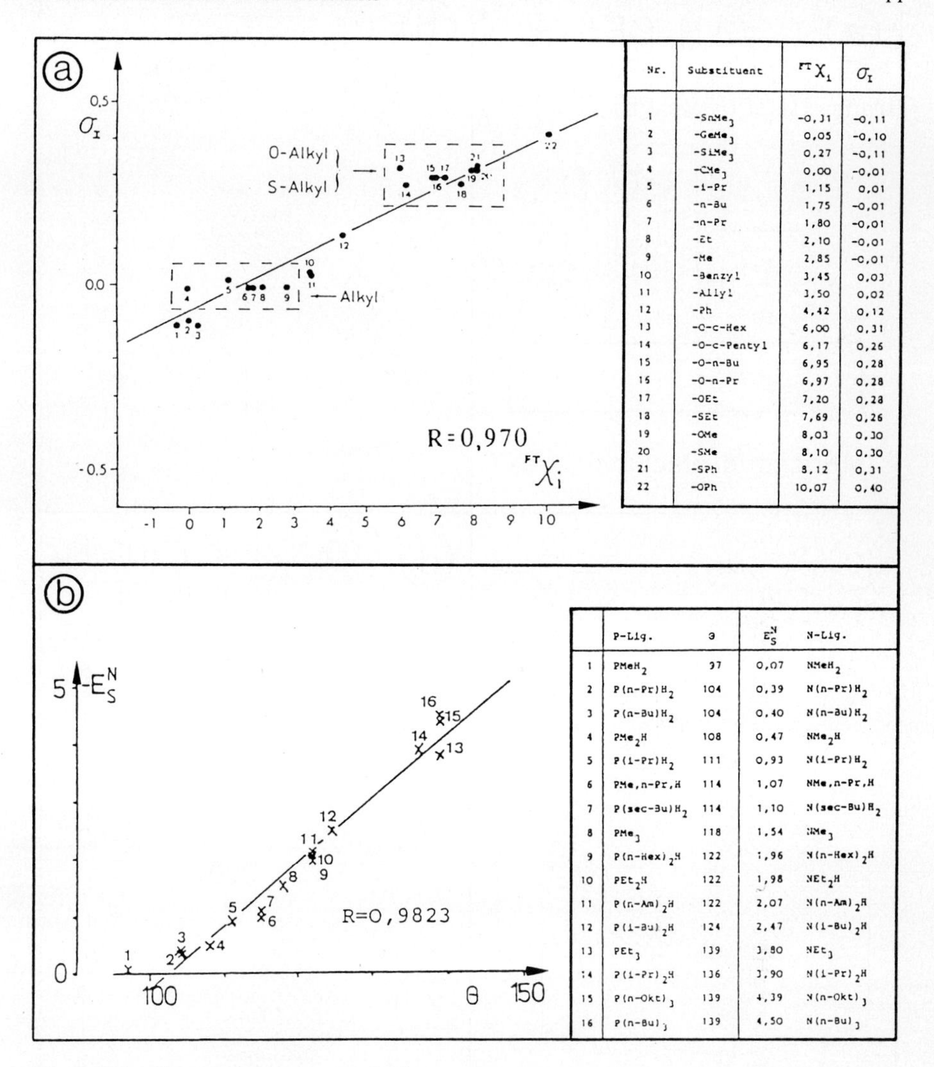

Nr.	Substituent	$^{FT}\chi_i$	σ_I
1	$-SnMe_3$	-0,31	-0,11
2	$-GeMe_3$	0,05	-0,10
3	$-SiMe_3$	0,27	-0,11
4	$-CMe_3$	0,00	-0,01
5	-i-Pr	1,15	0,01
6	-n-Bu	1,75	-0,01
7	-n-Pr	1,80	-0,01
8	-Et	2,10	-0,01
9	-Me	2,85	-0,01
10	-Benzyl	3,45	0,03
11	-Allyl	3,50	0,02
12	-Ph	4,42	0,12
13	-O-c-Hex	6,00	0,31
14	-O-c-Pentyl	6,17	0,26
15	-O-n-Bu	6,95	0,28
16	-O-n-Pr	6,97	0,28
17	-OEt	7,20	0,28
18	-SEt	7,69	0,26
19	-OMe	8,03	0,30
20	-SMe	8,10	0,30
21	-SPh	8,12	0,31
22	-OPh	10,07	0,40

	P-Lig.	θ	E_S^N	N-Lig.
1	$PMeH_2$	97	0,07	$NMeH_2$
2	$P(n\text{-}Pr)H_2$	104	0,39	$N(n\text{-}Pr)H_2$
3	$P(n\text{-}Bu)H_2$	104	0,40	$N(n\text{-}Bu)H_2$
4	PMe_2H	108	0,47	NMe_2H
5	$P(i\text{-}Pr)H_2$	111	0,93	$N(i\text{-}Pr)H_2$
6	PMe,n-Pr,H	114	1,07	NMe,n-Pr,H
7	$P(sec\text{-}Bu)H_2$	114	1,10	$N(sec\text{-}Bu)H_2$
8	PMe_3	118	1,54	NMe_3
9	$P(n\text{-}Hex)_2H$	122	1,96	$N(n\text{-}Hex)_2H$
10	PEt_2H	122	1,98	NEt_2H
11	$P(n\text{-}Am)_2H$	122	2,07	$N(n\text{-}Am)_2H$
12	$P(i\text{-}Bu)_2H$	124	2,47	$N(i\text{-}Bu)_2H$
13	PEt_3	139	3,80	NEt_3
14	$P(i\text{-}Pr)_2H$	136	3,90	$N(i\text{-}Pr)_2H$
15	$P(n\text{-}Okt)_3$	139	4,39	$N(n\text{-}Okt)_3$
16	$P(n\text{-}Bu)_3$	139	4,50	$N(n\text{-}Bu)_3$

Fig. 2.1a,b. Correlations of σ_I- and E_S^N-parameters for substituents of organic chemistry with Tolman's χ_i- and Θ_i-parameters of metalorganic chemistry

and the standard ligand $P(t\text{-}Bu)_3$ (2056,1 cm^{-1}). These parameters describe the influence of P-ligands on a metalorganic system whereby one of the two parameter sets may be sufficient depending on the type of system or both sets have to be applied. When both parameter sets meet the demands of determining the behavior sufficiently, this may be rationalized e.g. via the so-called "steric and electronic box" [56] (see Fig. 2.2.a).

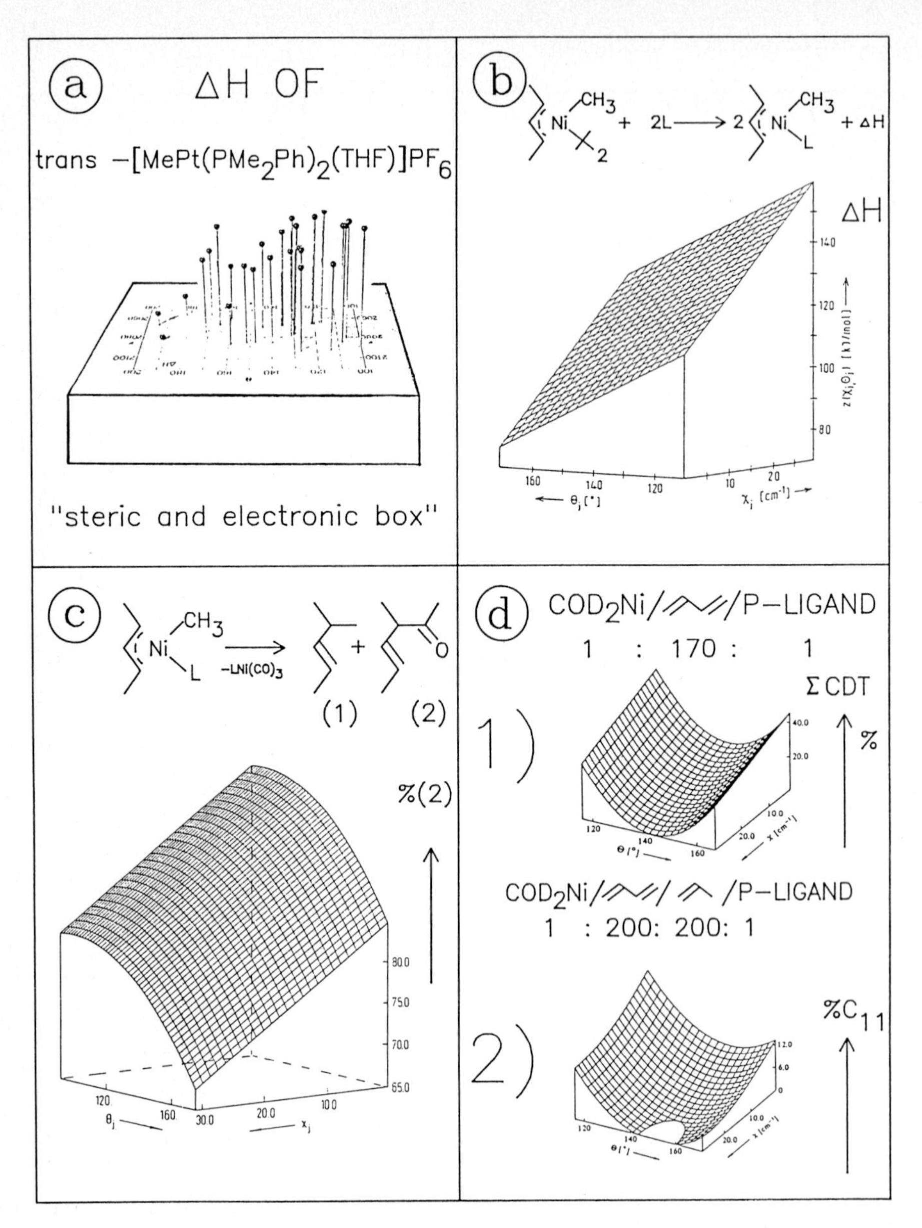

Fig. 2.2a. Enthalpy of reaction ΔH of *trans*-[MePt(PMe$_2$ Ph)$_2$-(THF)]PF$_6$ with an excess of L [60]. The height in cm is half the quantity given in kcals/mole: "steric and electronic box" [56]. **(b)** Presentation in form of a plane (profile of ligand property) of the dependency of the reaction enthalpies in the complexation of a ligand at a NI-complex on Tolman's parameters. For details see [61]. **(c)** Profile of ligand property for the portion of products of CO-insertion in the reaction with CO; for details see [62]. **(d)** Presentation as a hyper-plane of control by ligand property for the portion of cyclotrimers respectively 2:1 co-oligomers (two butadienes and one propene) in the three-components-system COD$_2$Ni/P-ligand/butadiene = 1:1:170 respectively in the four-components system COD$_2$Ni/P-ligand/butadiene/propene = 1:1:200:200 [20,64]

2.1 Correlation with the Parameter Sets Θ and X by Tolman

Four applications of Tolman's parameter sets Θ and X are summarized in Fig. 2.2. The influence of both parameter sets is demonstrated in the form of a "steric and electronic" box – according to Ref. [56] – for the reaction enthalpy of the association by an excess of L at a Pt-complex [60]. Referring to the example in Fig. 2.2d(1) scheme 2.2 casts light on the fact that no direct correlation between the individual parameter sets Θ and X may be derived from the data. The application of a multidimensional regression analysis results in a plane for the profile of ligand properties related to Θ- and X-parameter sets for the values of reaction enthalpy in this case of ligand association at a Ni-complex [61]. We got nonlinear planes of ligand control by regression analysis for the metal-induced CO-insertion between the 1,3-dimethyl-substituted allyl and the methyl group in Ni-complexes [20,62] and for ligand control of the percentages of cyclotrimers and 2:1-co-oligomers in the catalytic system Ni/L/butadiene (=1:1:170) respectively Ni/L/butadiene/propene (=1:1:200:200) in Fig. 2.2c and d [19,39,63,64].

But a disadvantage of the applied multidimensional regression analysis for the whole data set has to be seen by the fact that a relative simple function of the displayed type can not represent all possible forms of planes for a dependency of a behavior based upon two parameter sets (vide Scheme 2.3). Unsteady changes are not presentable at all in one function. A better fitting could be reached for a

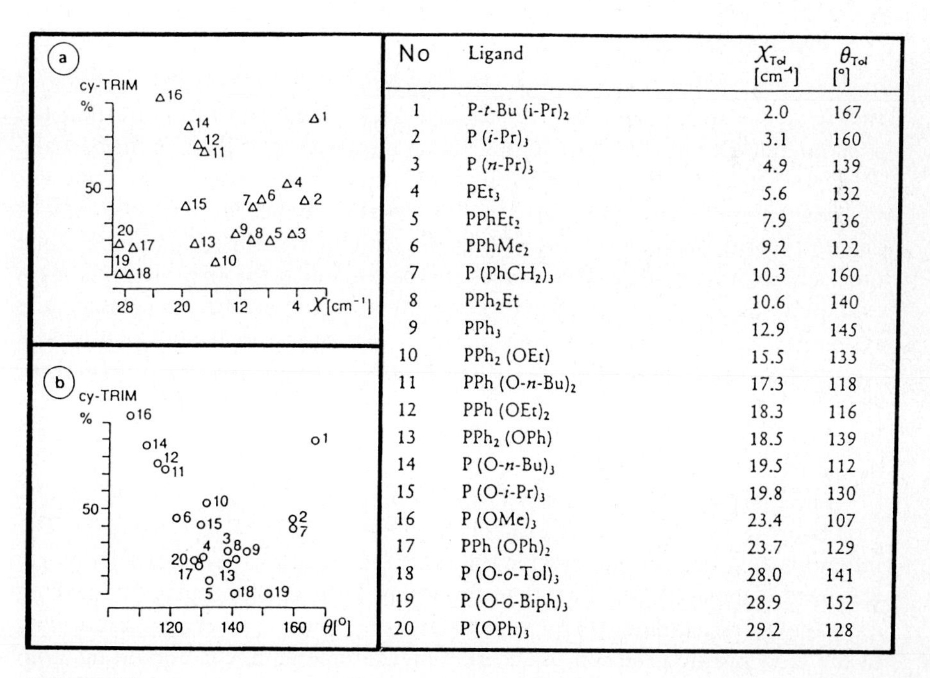

No	Ligand	X_{Tol} [cm^{-1}]	θ_{Tol} [°]
1	P-*t*-Bu (i-Pr)$_2$	2.0	167
2	P (*i*-Pr)$_3$	3.1	160
3	P (*n*-Pr)$_3$	4.9	139
4	PEt$_3$	5.6	132
5	PPhEt$_2$	7.9	136
6	PPhMe$_2$	9.2	122
7	P (PhCH$_2$)$_3$	10.3	160
8	PPh$_2$Et	10.6	140
9	PPh$_3$	12.9	145
10	PPh$_2$ (OEt)	15.5	133
11	PPh (O-*n*-Bu)$_2$	17.3	118
12	PPh (OEt)$_2$	18.3	116
13	PPh$_2$ (OPh)	18.5	139
14	P (O-*n*-Bu)$_3$	19.5	112
15	P (O-*i*-Pr)$_3$	19.8	130
16	P (OMe)$_3$	23.4	107
17	PPh (OPh)$_2$	23.7	129
18	P (O-o-Tol)$_3$	28.0	141
19	P (O-o-Biph)$_3$	28.9	152
20	P (OPh)$_3$	29.2	128

Scheme 2.2

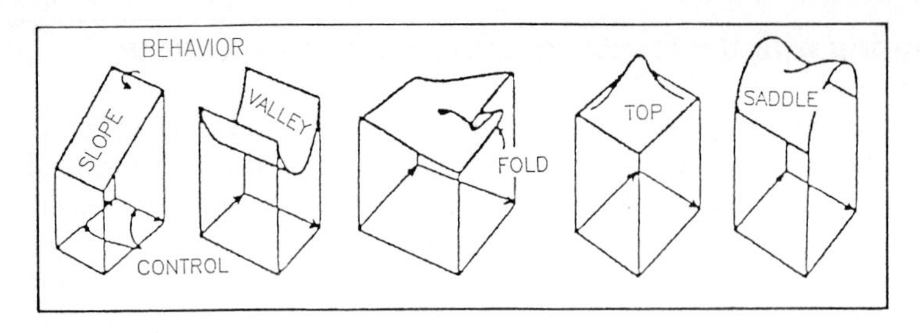

Scheme 2.3

sufficiently large data set by partial regression in small areas, followed by graphical combinations of the gained result. However, we have interrupted our efforts in this direction, because even the χ_i-parameters are not convenient for the characterization of the substituents. In addition we always have to neglect some data in the regression analysis, because they do not fit in (see e.g. the TPP in Fig. 2.8b).

We do not want to discuss further details of the "steric" parameter Θ here (see for this [65–67]). It can not be excluded e.g., that space filling influences the extent of s-character in the lone electron pair at phosphorus and thereby Θ perhaps becomes a criterion for the contribution of symmetric interactions.

2.2 Systematic Variations at Tetrahedral Centers: A New Parameter Set Δ

In Fig. 2.1 – in the squares with broken lines – one can recognize that the substituents have more differentiated values on the χ_i-scale than on the σ_I-scale. Therefore the exchange of three substituents for three others each time was investigated quite systematically. In Fig. 2.3 this defined exchange process is generally formulated for tetrahedral centers with the coordination number three or four and four investigated examples are represented in (b). In principle, always four quantities can be expected in investigations of structures or processes by such systematic variations. In Fig. 2.4a and b, it is demonstrated by which clearly separable patterns these quantities are related. Nevertheless mixed forms of these patterns may be observed in reality. In (c), it is shown which superposed patterns may be expected e.g. arising from collective data. We will discuss the multidual decision trees in detail later on (see Chap. 5).

The sharp ν_{co} (A_1)-band of $(CO)_3NiL$-complexes – proposed by Tolman – may be measured much more exactly by means of a FT–IR-spectrometer. The inherent calibration of the wavelength inside the apparatus (Connes' advantage) was used extensively for exact determination. Tests of reproducibility gave an error for the measurement of the wavelength of $\pm 0{,}1\ cm^{-1}$. A very good additivity of the data resulted even in these exact measurements in corresponding complexes when three substituents were replaced by three others at the phosphorus

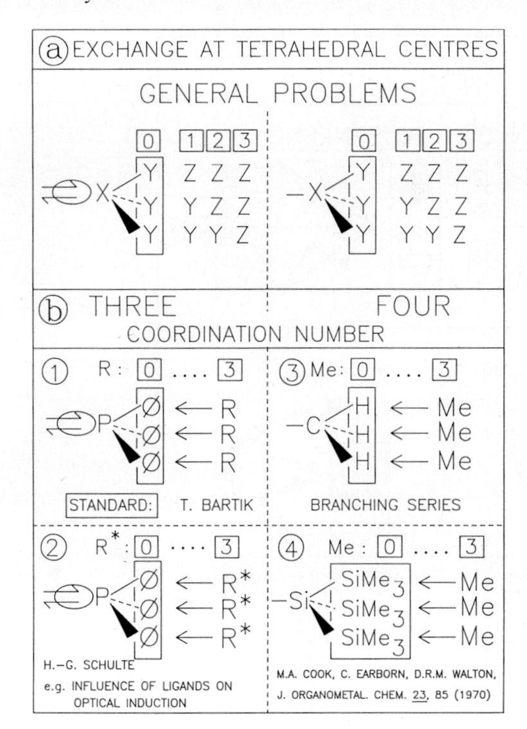

Fig. 2.3a. Systematic exchange of one type of substituents by another at tetrahedral centers with the coordination number three or four. **b** Four real examples

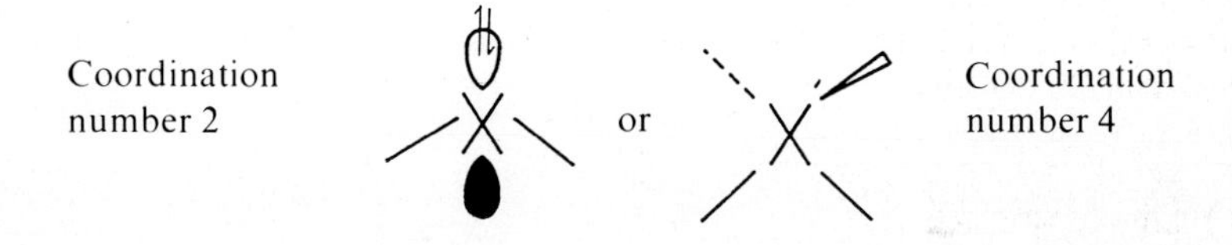

(e.g. methyl- against t-butylgroups). But starting from TPP and replacing step by step the phenylgroups by those substituents X with central atoms of the fourth or sixth main group (restriction for coherency is the EVEN coordination number two or four) patterns are observed with overwhelmingly monotonous deviation from linearity (vide Fig. 2.5). When the averaged values Δ of the deviations from linearity belonging to the maxima of the υ_{co} (A_1)-data of Ph_2XP– and PhX_2P–Ni–complexes are plotted against the P-parameters of the substituents – determined via PX_3–$Ni(CO)_3$-complexes – the meta-pattern of Fig. 2.6a occurs. In Fig. 2.6b the multifold meta-pattern of data are represented in addition by Me/Et/i–Pr/t–Bu variation in alkyl- and alkoxi-substituents of Ph_2XP– (black symbols) and PhX_2 P-compounds (white symbols) [59,68,69]. See also data sets in the appendix at the end of the book.

The alkyl- (+ +), alkoxi- (——), thio-alkoxi- (— +) and EMe_3-substituents (+ —) are precisely identified as classes by the positive and negative signs of the Δ- and P-parameters. Triphenyl-phosphane is surely a very appropriated standard because it exists as a degenerated racemic 1:1 mixture of conformations with

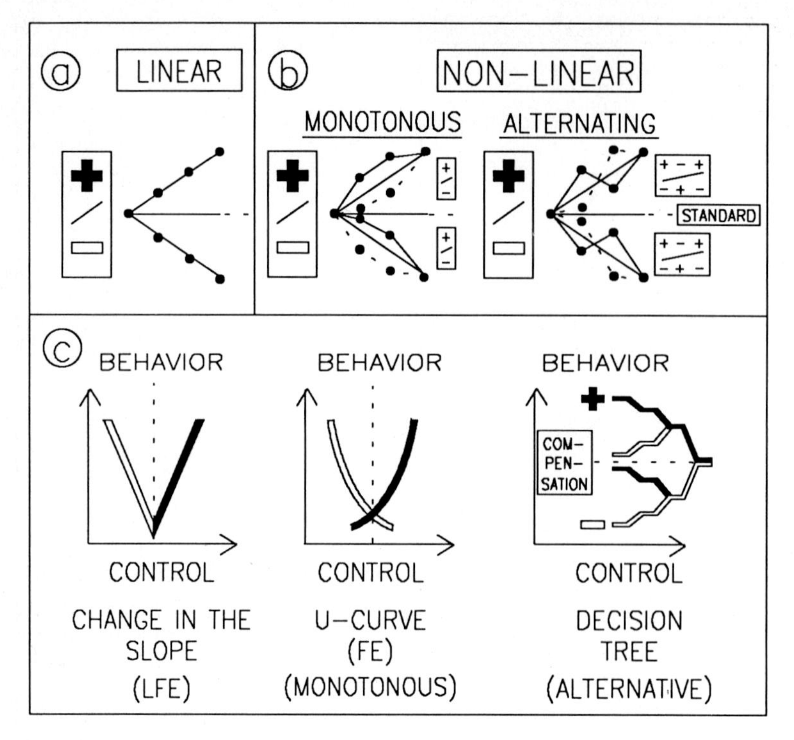

Fig. 2.4a-c. Patterns derived from four quantities: 2 × LINEAR in **(a)**, 4 × MONOTONOUS and 4 × ALTERNATING deviating in **(b)**. In **(c)** the expected patterns of collective data are depicted

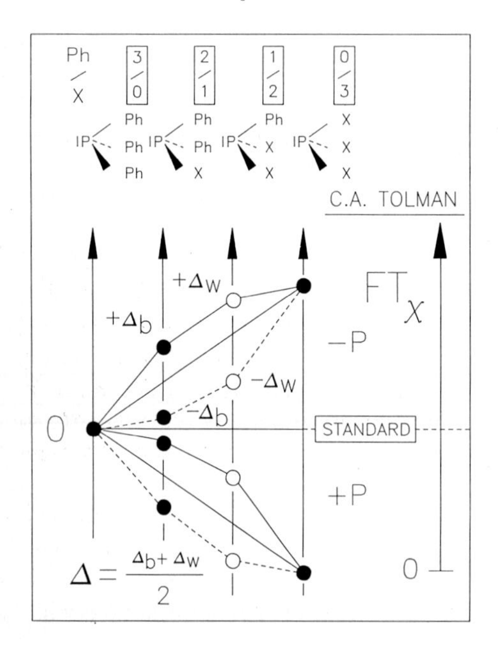

$$\Delta = \frac{\Delta_b + \Delta_w}{2}$$

Fig. 2.5. The additivity rule of Tolman is not fulfilled in the represented investigations. Only MONOTONOUS deviating patterns are obtained, when the phenyl groups of the triphenylphosphane (new standard) are replaced by substituents with central atoms only from fourth or sixth main group (EVEN number of valence electrons)

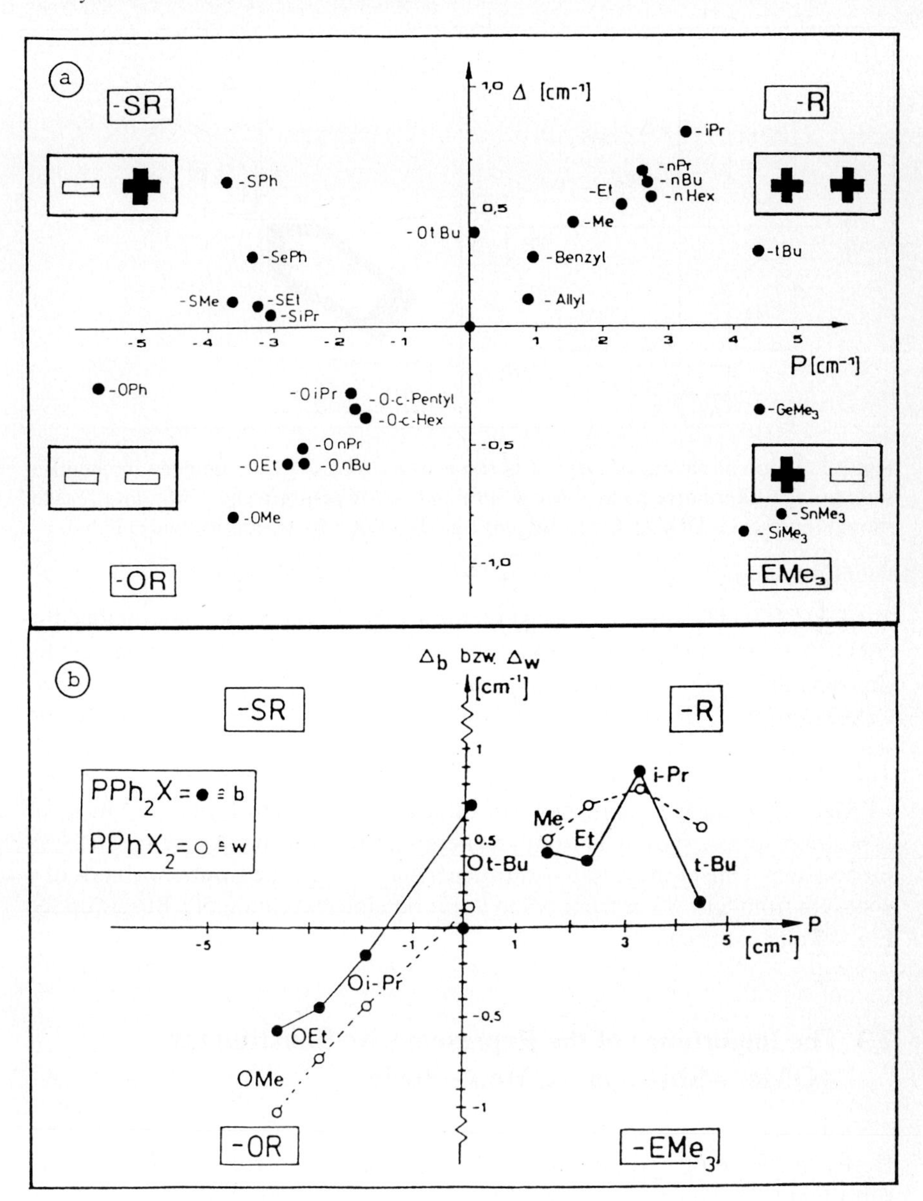

Fig. 2.6a. Plotting the MONOTONOUS averaged deviations $\pm\Delta$ versus the $\pm$ P-parameters results in a meta-pattern, in which the alkyls, alkoxides, thio-alkyls and EMe_3-substituents are classified via the two signs of Δ and P. **(b)** Different patterns from four quantities of the alkyl and alkoxide series are obtained by systematic variation of the alkyls at the tetrahedral C-atom (Me/Et/i-Pr/t-Bu)

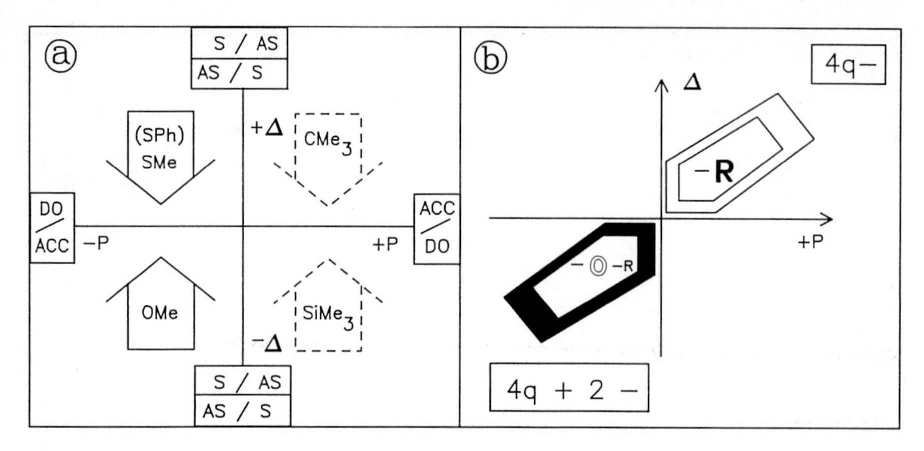

Fig. 2.7. Proofs in favour of DO/ACC respectively ACC/DO-phenomena in chemical systems. The alternative pairs -SMe (-SPh) and -OMe respectively $-CMe_3$ and $-SiMe_3$ should represent a DO/ACC control and e.g. -R/-OR a S/AS control; vide [28,59]

"LEFT/RIGHT" propeller arrangements. In Fig. 2.7 [28,59] we have named the ORDER FACTORS, which perhaps determine the signs of the parameters by choosing the type of central atoms in the substituents. O/S show *absolutely* and C/Si *relatively* DO/ACC-behavior towards electrons – always related to the standard carbon – being derived from the first ionization potential (vide Fig. 3.1). The substituents –SR and –OR are to be distinguished – compared to $-CR_3$ and $-EMe_3$ – by the fact, that they can offer a lone π-electron pair e.g. in α- or ω-position to a π-system of similar energy and symmetry correctness (S/AS subsystems). One of these two substituents may change the frontier orbitals of a π-system from S to AS or from AS to S (for possible exception of *t*-Bu-group see Fig. 4.7 and context).

2.3 The Importance of the Representative Substituents –OMe/–SMe and $-CMe_3/-SiMe_3$

Here some examples will be used to demonstrate that the different families of substituents classified by the signs of Δ and P are able to effect inverse order in structures and processes of parent systems by their alternatives DO/ACC respectively AS/S of their central atoms. For this we have chosen the ACC/DO-pairs in the central atoms of –OMe/–SMe and $-CMe_3/-SiMe_3$ as representative substituents. It is important thereby to notice that these pairs do not differ in all cases in the signs (and quantity) of the P-parameters – which may be used instead of σ_I-parameters – but their Δ-parameters have opposite signs and nearly the same value. –R/–OR substituents e.g. differ by the signs of both parameters (vide appendix Scheme 11.1).

In Fig. 2.8a the inverse influence of –R*/–OR* is demonstrated on the enantiomeric excess formed in dependence on the amount of correspondingly substituted P-ligands (for the discontinuous method of INVERSE TITRATION see Chap. 6) referring only to one example of optical induction in the metal-induced synthesis of 3-methyl-E-4-hexen-2-one [70,71]. The inverse influence in every case of alkyl- and alkoxi-substituted P-ligands is depicted in Fig. 2.8b for the selectivity of eight-membered rings among the cyclodimers from butadienes in the ligand modified Ni-catalysis in dependence on the quantity – but especially on the sign – of P-parameters [72].

In Fig. 2.9a the effect of the representative substituents $-OCH_3$/$-SCH_3$ and $-CMe_3$/$-SiMe_3$ in the 2-position of 1,3-dienes is shown for the quantities of ^{13}C-shift-data e.g. of the C1- or C4-atoms in these 1,3-dienes [73] with black/white arrows in the same especially clearcut way as in Fig. 4.1 and Scheme 4.1. The absolute/relative influences of representative substituents are dependent both on the system as well as on the position in the system and on the type of investigated system properties. This is demonstrated by a comparison of patterns presented in (a) with those of the ionisation energies of the two highest π-MO's of the correspondingly 1,4-substituted benzene-derivatives in Fig. 2.9b [68,74]. Depending on the models shown in Scheme 2.4 [74] 1,4- or 2,5-dihydro-benzene-derivatives are formed, which is determined by the symmetry of the frontier molecular orbitals of the arenes. The influence of –OMe/–SMe in 1,4-positions of benzene-derivatives is so decisive, that a totally different reaction is observed [68] in the case of sulfur-derivatives – namely the splitting off by the aliphatic side chain coupled to the sulfur (vide Fig. 2.10a). The influence exerted by O/S-heteroatoms as couplers in alkyl-aryl-systems on the reactivity in the nucleus or the aliphatic side chain is not a unique observation, e.g.

$$LiCH_2{-}S{-}C_6H_5 \xleftarrow[-)n\text{-}Bu\text{-}H]{} CH_3{-}S{-}C_6H_5 \;/\; C_6H_5{-}O{-}CH_3 \xrightarrow[-)n\text{-}Bu\text{-}H]{} (o\text{-}Li)C_6H_4{-}O{-}CH_3 \qquad + \text{Li-}n\text{-Bu}$$

Even the 1,4-trimethylsilyl- and *tert*-butyl-benzene-derivatives may be compared in their 1,4–/2,5-regioselectivity only relatively. As shown by Scheme 2.5 the reduction to 2,5-derivatives of 1,4-dialkylsubstituted benzene derivatives decreases dramatically with the degree of branching [68].

1,3-dienes synthesized with representative substituents in 2-position form generally Diels-Alder products with acrylic acid-Me/Et/i-Pr/t-Bu-esters. The known relative reactivities from competition experiments with 2-phenyl-buta-1,3-diene as the chosen standard in Fig. 2.10c and the regioselectivities found (% "*para*" + % "*ortho*" $\approx$ 100 %) in Fig. 2.10d are easily recognized in the form of patterns in absolute/relative control [73].

In Fig. 2.10b, an extremely interesting case of the influence via the alternatives AS/S and ACC/DO in 2-position of representatively substituted 1,3-dienes is described on the stabilization/destabilization of educt-/intermediate-/product-complexes [73,75]. The exchange of products under formation of educt-complexes and so on corresponds to catalytic processes in cycles of the

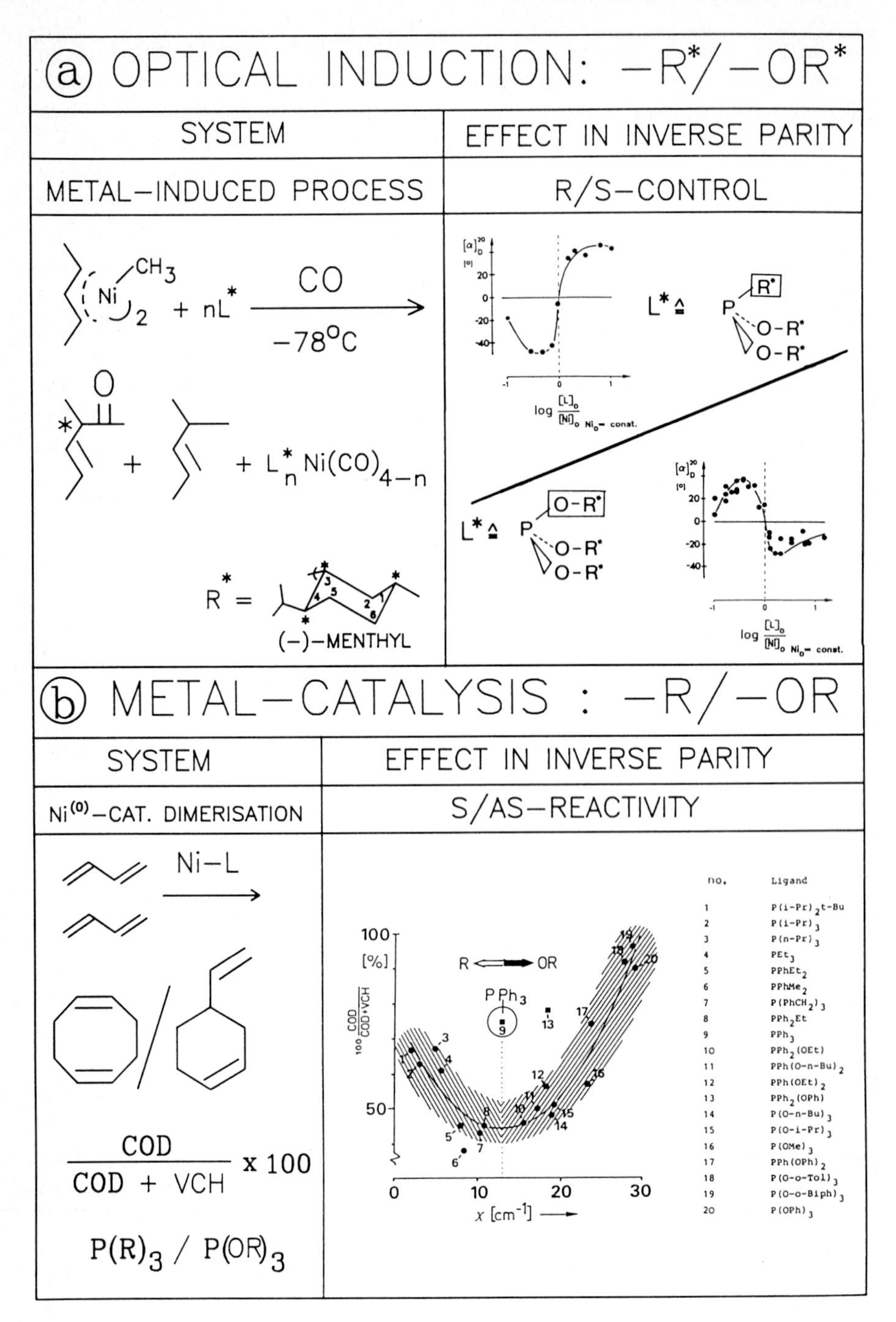

no.	Ligand
1	$P(i\text{-}Pr)_2t\text{-}Bu$
2	$P(i\text{-}Pr)_3$
3	$P(n\text{-}Pr)_3$
4	PEt_3
5	$PPhEt_2$
6	$PPhMe_2$
7	$P(PhCH_2)_3$
8	PPh_2Et
9	PPh_3
10	$PPh_2(OEt)$
11	$PPh(O\text{-}n\text{-}Bu)_2$
12	$PPh(OEt)_2$
13	$PPh_2(OPh)$
14	$P(O\text{-}n\text{-}Bu)_3$
15	$P(O\text{-}i\text{-}Pr)_3$
16	$P(OMe)_3$
17	$PPh(OPh)_2$
18	$P(O\text{-}o\text{-}Tol)_3$
19	$P(O\text{-}o\text{-}Biph)_3$
20	$P(OPh)_3$

Fig. 2.8a,b. Alternative effects of −R*/−OR* substituents on the direction of optical induction [71] in **(a)** and of −R/−OR on the chemoselectivity of a 4q−/4q+2-system in **(b)** [72]

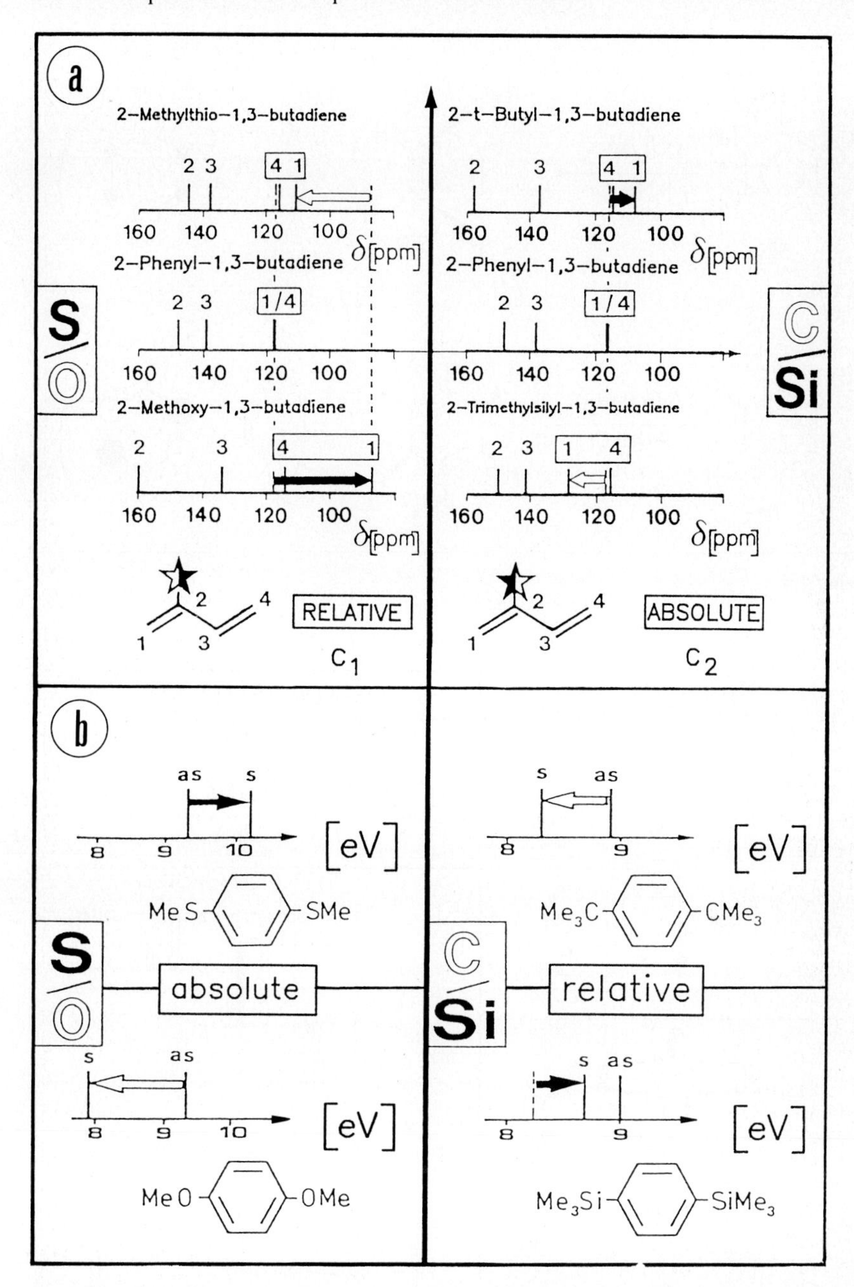

Fig. 2.9a,b. Influence of the representative substituents -OMe/-SMe and $-CMe_3$/ $-SiMe_3$ on the patterns of certain parameters in ^{13}C-NMR in **(a)** [73] and in photoelectron spectroscopy in **(b)** [68] (tentative assignment for C4, C1 in S-Me-derivative)

SiR3 H H H H SiR3

(Na/NH_3) $+NH_4Cl$

R3Si H H H H H H SiR3

$R \triangleq CH_3$

π_s

CH3 H H H H CH3

(Na/NH_3) $+NH_4Cl$

CH3 H H H H H H CH3

π_{as}

Scheme 2.4

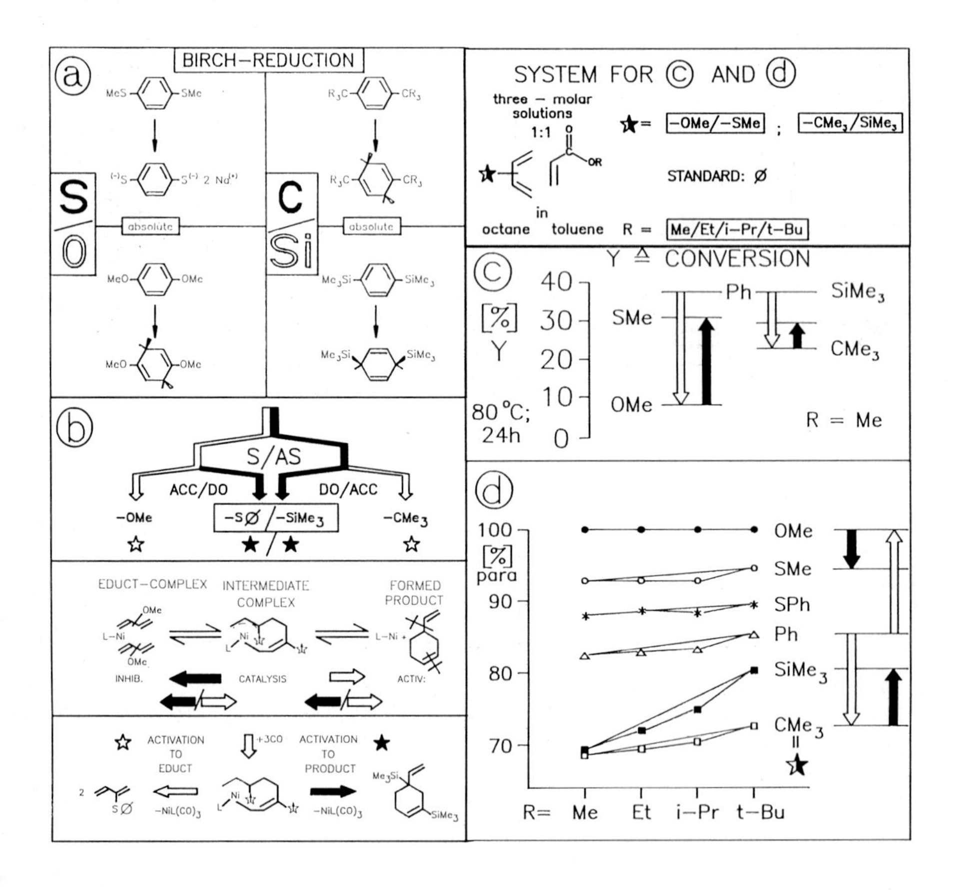

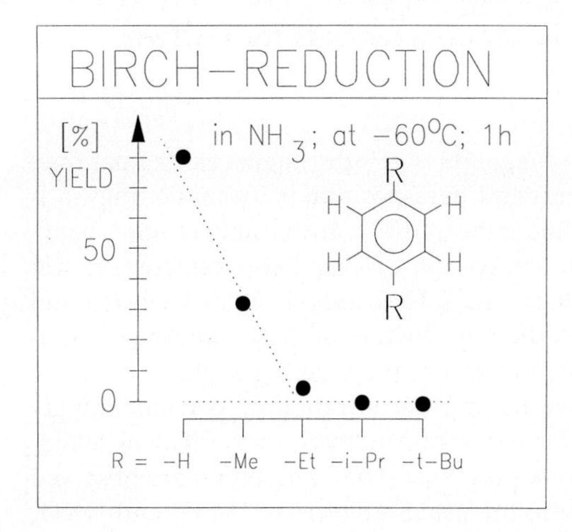

Scheme 2.5

$Ni^{(0)}$/L/butadiene system. At ligand-free $Ni^{(0)}$-catalysts both in 2-position trimethylsilyl- and *tertiary*-butyl-substituted 1,3-dienes react faster than butadiene to six-membered ring derivatives [76]. (The 2-trimethylsilyl-buta-1,3-diene is converted a hundred times faster than 2-*tertiary*-butyl-1,3-diene in a thermal Diels-Alder reaction.) Indeed the addition of one mole TPP strongly inhibits the catalytic formation of six ring dimers from 2-*tertiary*-butyl-1,3-diene and changes the distribution of regioisomers [76], but the catalytic synthesis of six-membered rings from 2-trimethylsilyl-1,3-diene is totally suppressed and only a corresponding eight-membered chain at the Ni•TPP is spectroscopically detectable [75]. While only bis-1,3-diene-Ni-TPP-complexes can be recognized by ^{13}C-NMR-spectroscopy when 2-methoxy-1,3-diene reacts with Ni $(COD)_2$ and TPP (1:1), the 2-thiophenoxy-1,3-diene leads only to the formation of a C_8-chain at the Ni-TPP [73].

Interestingly, the two C_8-chains at Ni-TPP – with nearly identical structures – behave inversely when reacting with CO. C-C-splitting under educt formation occurs in the "S"-complex and C-C-coupling to a six-membered ring derivative is observed in the "Si"-complex [73,75].

Fig. 2.10a-d. Influence of the representative substituents -OMe/ -SMe and $-CMe_3$/ $-SiMe_3$ on the reactivities of chemical systems: **(a)** Influence on the regio-selectivity in the Birch reduction [68]; **(b)** Influence on activation/inhibition phenomena in a $Ni^{(0)}$/TPP/2-substituted 1,3-diene system on a reaction cascade [73,75] (vide text). **(c)** and **(d)** Influence on the conversion (relative reaction rates in competition) and on the regioselectivity in the Diels-Alder reaction [73]

2.4 The Importance of P- and Δ-parameter Sets for Pattern Comparison

While it is assumed that the controlling effects of substituents can be parameterized relatively easily for the meta- and para-position in aromatic rings, it is accepted, that the consequences for the ortho-position are harder to comprehend. We have tried to authenticate the application of P-and Δ-parameter sets for selectivities by pattern comparison in Fig. 2.11 for alkyl- (a), alkoxi- (b) and dialkylamino-substituents (c). With the introduction of these pattern comparisons we leave the level of just tabulating quantities (vide Fig. 4.7).

For this purpose we have investigated the nitration of correspondingly substituted benzene-derivatives under coherent conditions (only NO_2^+) by avoiding redox processes [14]. These redox processes (NO^+ and NO_2^+) are observed – shown by the brown red gases – in the usual mixtures of H_2SO_4 and HNO_3 (mixed acid). Thereby information on regioselectivity disappears, because the nitroso compounds with their inverse rules of regioselectivity are oxidized to nitro derivatives. The phenyl group is chosen as the standard for the parameter sets. By plotting for pattern comparison instead of the Δ_b-parameters of the PPh_2X-compounds – the negative logarithm of the *ortho-*/2×*para*-percentages against the P-scale – an astonishing similarity of the patterns is observed after adjusting the scales. This is even true for "exceptions" such as -O-*t*-Bu and -N(*i*-Pr)$_2$ (see Fig. 2.11). It is worth mentioning that we can totally give up all claim to "steric" parameters (see Fig. 3.23 and context).

2.5 The Importance of Compensation Phenomena

The very strong differences in the pairs of representative substituents -OMe/-SMe and -CMe_3/-$SiMe_3$, which we described in Sect. 2.3, fundamentally contradict the fact that these pairs of substituents have practically the same σ_I-parameters, which are used with great success in FE- and LFE-relations. Can this apparent contradiction be understood? Yes! Compensation phenomena (vide Fig. 2.12) explain a lot of things and we should be aware of the fact, that in every case the described parameter sets represent *the substituents and the couplers,* which overcome the introduced inverse perturbations by choosing inverse conformations via self-organization. How to prove this by experiments, we are still investigating by series of X-ray determinations.

We would like to show by two examples – comparing determinations of parameters and regioselectivities – that there are simple possibilities of verifying these compensations by experiments. When determining the parameter sets for X-substituents by successive replacement of all phenyl groups in triphenylphosphane a compensating coupler in every case is inserted between the phosphorus and -X. The result – exemplified by the representative substituents -OCH_3/-SCH_3 and -CMe_3/-$SiMe_3$ supplying -CH_2- (methylene group) or *para-*

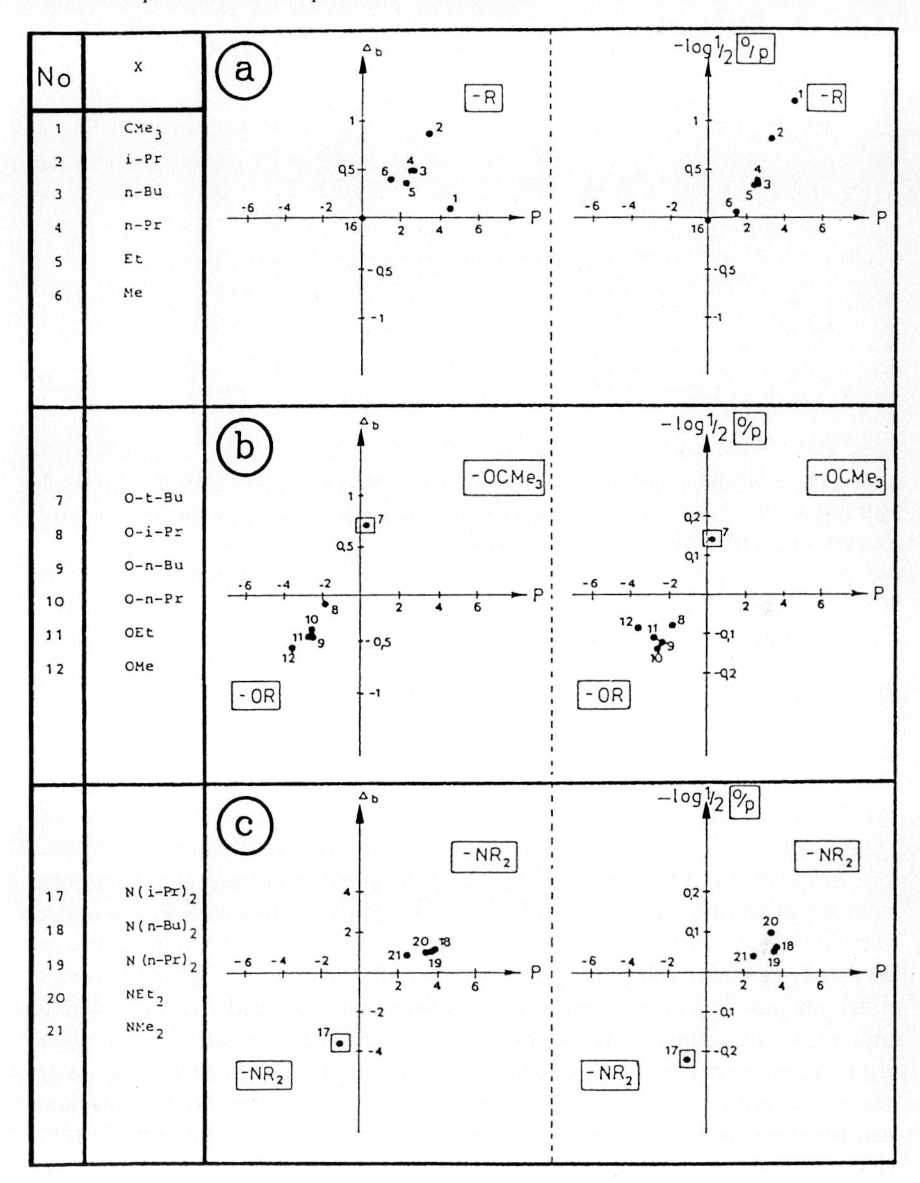

Fig. 2.11. Comparison of the patterns of the Δ_b parameters versus the-log o-/2 p-quantities obtained in the nitration under coherent conditions of correspondingly monosubstituted arenes plotted in every case against the P-parameter sets [14] (standard ≙ phenyl ≙ 16)

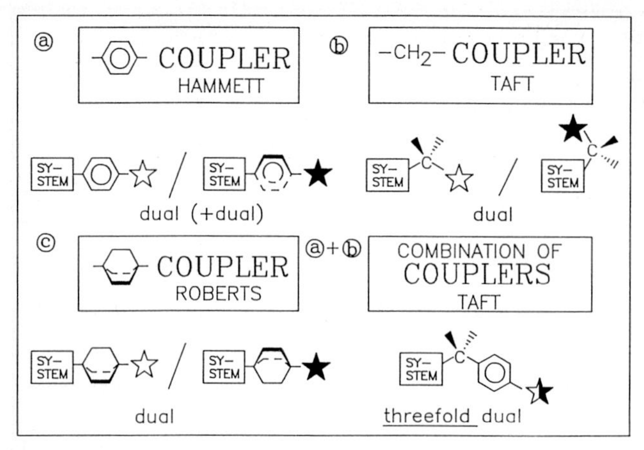

Fig. 2.12. The individual substituent with ALTERNATIVE PRINCIPLES induces, by self organization, alternative conformations of the system. The "parameters of substituents" characterize substituents *and* couplers

phenylene as couplers – is depicted in Fig. 2.13a. The P-parameters are essentially less differentiated and a deviation from additivity, which can be easily recognized by omission of the Δ-parameters, is only caused by $-CH_2-SiMe_3$; its Δ-value has had its sign changed, too [68,69].

The elimination of inverse effects – caused by different ALTERNATIVES ACC/DO in the central atoms of substituents such as -OMe/ -SMe and $-CMe_3/-SiMe_3$ – can be proved by pattern comparison of the stereoselectivities e.g. in the Wittig reaction (vide Fig. 2.13b) by self-organized conformational compensations in the couplers. Substituting the ylide position with representative subsystems such as -OMe/-SMe and $-CMe_3/-SiMe_3$ and varying, at the same time in the series under coherent conditions, the alkyl groups Me/Et/*n*-Pr/*n*-Bu in aldehyde position, inverse patterns of *cis*-selectivity are observed for both pairs and – by comparing $-CMe_3/-SiMe_3$ – one even recognizes a change in *cis*- to *trans*-stereoselectivity. But inserting in the individual series in every case a *para*-phenylene group between the ylide position and the representative substituents results in an elimination of all differences in patterns by compensations. While the

Fig. 2.13a. The insertion of methylene or *para*-phenylene groups between phosphorus and the representative substituents – when determining Δ- and P-parameters – results in far reaching compensations of P-parameters and nearly total compensation on the Δ-scale. **(b)** The differentiated patterns of cis-selectivity – by Me/Et/n-Pr/n-Bu variation in aldehyde and variation by representative substituents in ylide position of a Wittig system – show a totally uniform picture by compensations after insertion of a *para*-phenylene group at the ylide position [31]

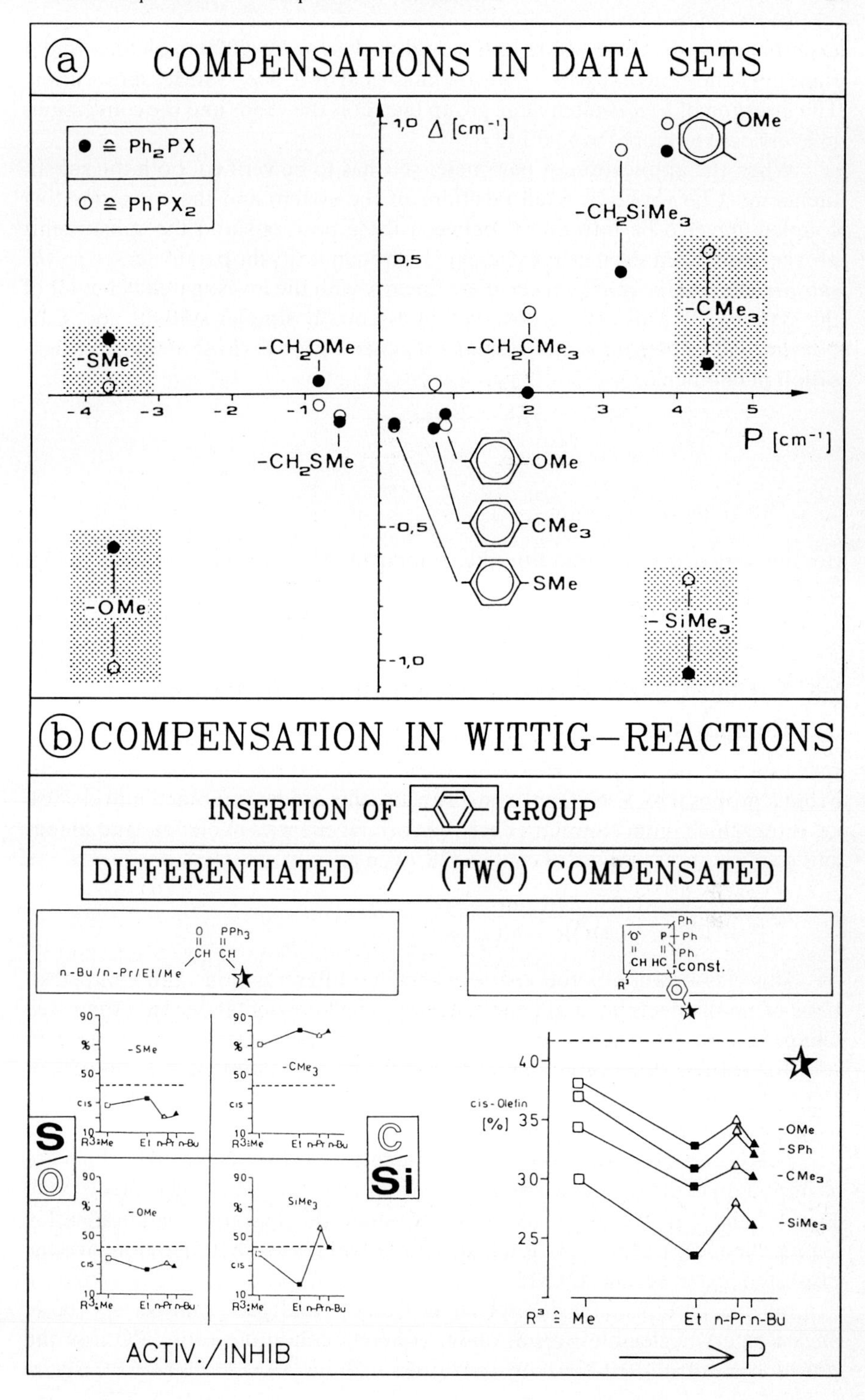

(a) COMPENSATIONS IN DATA SETS
● ≙ Ph2PX
○ ≙ PhPX2
Δ [cm⁻¹]
P [cm⁻¹]
OMe
-CH2SiMe3
-CMe3
-SMe
-CH2OMe
-CH2CMe3
-CH2SMe
OMe
CMe3
SMe
-OMe
-SiMe3
(b) COMPENSATION IN WITTIG-REACTIONS
INSERTION OF GROUP
DIFFERENTIATED / (TWO) COMPENSATED
n-Bu/n-Pr/Et/Me
PPh3
const.
-SMe
-CMe3
-OMe
SiMe3
S
O
C
Si
cis-Olefin [%]
-OMe
-SPh
-CMe3
-SiMe3
R3 ≙ Me Et n-Pr n-Bu
ACTIV./INHIB
P

experiments with -OMe/-SMe lead to high yields ($\approx$ 70–90%) in olefines under standard conditions, only 10–15% in olefines yield by -CMe_3/-$SiMe_3$ substitution. The insertion of a *para*-phenylene group increases the yields and the conversions in every case to more than 90% [31].

When the application of parameter sets has to be verified, both the substituents must be varied at fixed positions of the system and the corresponding couplers have to be introduced between these positions and the substituents in every case. Then, even in biosystems of high complexity the parameter sets (*of the substituents and the couplers*) correlate linearly with the investigated behavior of the system [77]. This can be observed in essentially simpler systems, too. Carboxylic acid ester groups and similar subsystems always cause severe compensation phenomena.

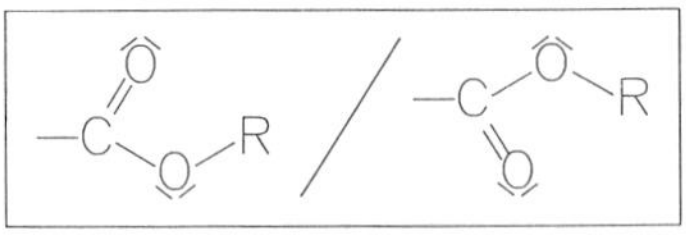

(for this see e.g. the nearly additive behavior in the Me/Et/*i*-Pr/*t*-Bu series in Fig. 2.10c and d).

2.6 Further Characterization of Substituents by Pattern Comparison

From a proposal by K.N. Houk some substituents can be combined into classes, i.e. those which induce similar effects, e.g. orbital energies of olefines and dienes and coefficients at individual centers [78,79], e.g.

$-\ddot{X} = -R, -OR, -NR_2$ etc.
$-Z = -CN, -COOR, -NO_2$ etc.

This classification is too coarse when the differentiations and compensations of small effects have to be investigated in whole digital decision trees (see Chap. 5).

For further differentiations of "DO"/"ACC" substituents we suggest in Fig. 2.14, the EVEN/ODD number of valence electrons of heteroatoms in α-/β-position of these substituents.

This is derived from a series of substituents – H/CH_3/NH_2/OH, which are so important in the various carriers of information in the genes and proteins (vide Figs. 9.2 and 9.3). We would like to demonstrate this has great significance for simple chemical systems, too. But first of all we will start with simple comparisons of isosteric pairs such as O/NH.

Inverse patterns of *ortho*-selectivity (*ortho*- and *para*-isomers = 100%; *meta*-isomers neglegible!) result under coherent conditions (only NO_2^+) in the nitration of substituted benzene derivatives with the alkyl series Me/Et/*i*-Pr/*t*-

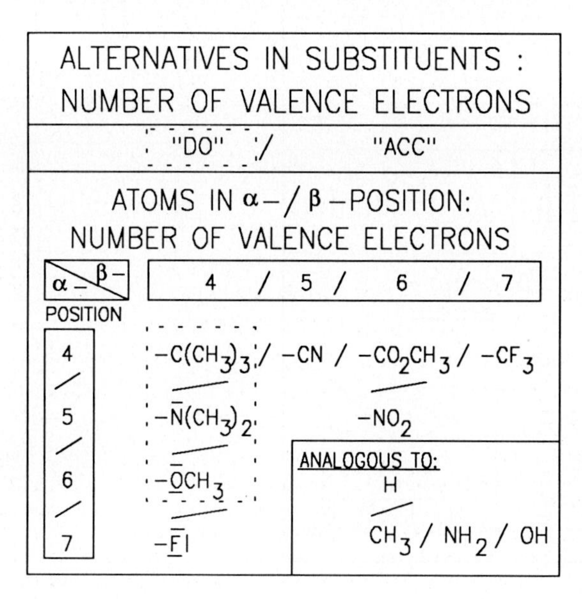

Fig. 2.14. Systematic variations in "DO"/"ACC" substituents [23] by the choice of heteroatoms with EVEN/ODD number of valence electrons in α-respectively β-position (for this series $H/CH_3/NH_2/OH$ vide Fig. 9.2 and 3)

Bu, which are coupled to the arenes by the isosteric couplers O/NH [14] (vide Fig. 2.15). The quantity of *ortho*-isomers decreases by space filling in the case of alkoxi substituents, while the opposite trend is realized with the alkyl-anilines. We obtained more than 78% of the *ortho*-nitro-derivative with di-isopropyl-aniline (for relative reaction rates see Fig. 2.16b). The reaction mechanism of nitration should be qualitatively different here from the course of the reaction with the alkoxi derivatives (SET following [80]).

The space filling of substituents is surely overestimated as an argument for selectivity control (vide [81]). For example, Si = Si double bonds are surrounded "by concrete" with the help of *tertiary*-butyl groups [82] but are presented in the same review a short time later as highly reactive systems e.g. for the synthesis of three-membered rings. Here we have to take into consideration the exceptional situation of the *tertiary*-butyl group with a partial positive charge [83] and the enhanced collapse of σ-/π-separation with the increasing number of identical alkyl subsystems. The additional change of partial charge at the central atom to a positive sign – in contrast to the negative sign in all other alkyl groups – motivated us, to present cautiously the branching series according to ALTERNATIVE PRINCIPLES by the following symbols:

Me/Et/*i*-Pr//*t*-Bu.

Four further examples of absolute inverse behavior in "O/N"-systems are summarized in Fig. 2.16. In (a) the choice of $-OEt/-NEt_2$ as substituents leads to totally different consecutive reactions after a 1,3-dipolar cycloaddition [84,85]. In (b) the alternation of relative reaction rates in nitration – by choosing the atoms C/N/O with EVEN/ODD/EVEN numbers of valence electrons as central atom

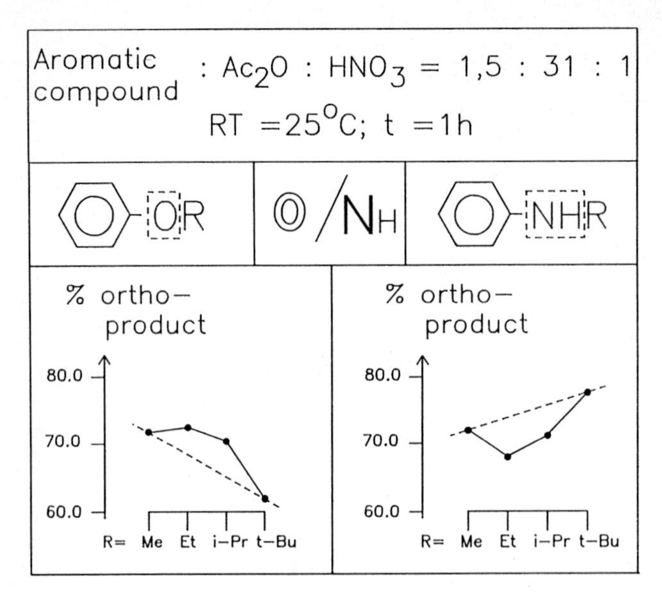

Fig. 2.15. Influence of isosteric couplers – with EVEN/ODD numbers of valence electrons at the central atom – on the pattern of *ortho*-selectivity in the nitration of monosubstituted arenes with identical alkyl variation [14]

of the couplers of two aromatic systems – is contrasted by comparison with the alternation caused by the number of structural units -CH_2-within the couplers (0,1,2) [14]. In (c) alkyl variations (Me/Et/*i*-Pr/*t*-Bu) in α-amino- or α-hydroxi-carboxylic acids exercise inverse influence on the extent of optical induction in the asymmetric synthesis of alcohols as modifying reagents in the hydrogenation with Raney-Ni-catalysts [86]. The allylation by isopropoxi- or dimethylamino-Ti-compounds in (d) is selective for ketones with an OPEN (aliphatic) or CYCLIC structure in a competitive situation [87] (Attention: the number of subsystems -CH_2- varies, too, but is surely lower in the hierarchical order!). For further examples see Fig. 5.3.

→

Fig. 2.16a. Different influence of alkoxy- and dialkylamino-substituents on the consecutive reaction from the reaction of a 2,3-diaza-1,3-diene derivative with ethoxy and diethylamino-substituted propyne derivatives [84,85]. **(b)** Relative reaction rates of α, ω-diphenyl-derivatives-standard = 1 for biphenyl-depending on number of methylene groups or type of isoelectronic couplers. **(c)** Influence of a hydroxy- or amino-substituent in a modifying reagent on the enantioselectivity in the hydrogenation of an acetoxy acetic acid methylester applying modified Raney-Ni-catalysts [86]. **(d)** Selective allylation of carbonyl compounds with allyl-Ti-alkoxides or -amides [87] (*)vide text

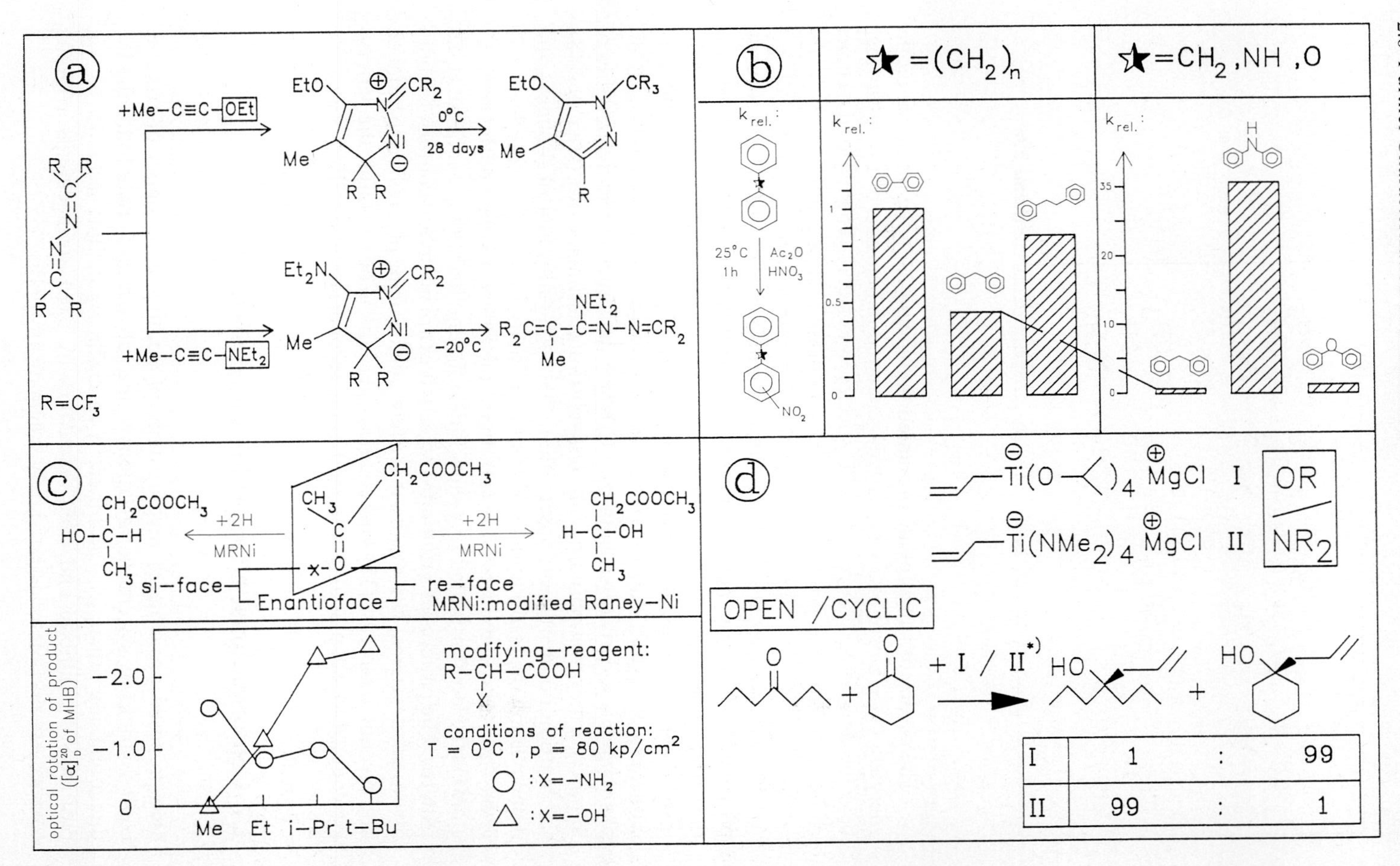
a
+Me–C≡C–OEt
0°C
28 days
+Me–C≡C–NEt$_2$
–20°C
R=CF$_3$
b
k$_{rel.}$:
25°C
1h
Ac$_2$O
HNO$_3$
NO$_2$
★=(CH$_2$)$_n$
★=CH$_2$,NH,O
c
+2H
MRNi
si–face
Enantioface
re–face
MRNi:modified Raney–Ni
optical rotation of product
([α]$^{20}_{D}$ of MHB)
Me Et i–Pr t–Bu
modifying–reagent:
R–CH–COOH
X
conditions of reaction:
T = 0°C , p = 80 kp/cm^2
:X=–NH$_2$
:X=–OH
d
MgCl I
MgCl II
OR
NR$_2$
OPEN /CYCLIC
+ I / II*)
I 1 : 99
II 99 : 1

Scheme 2.6

The classes of substituents presented in Scheme 2.6 have influence on the energies of AS/S orbitals, which are degenerated in benzene, of 1,4-substituted benzene-derivatives compared by PE-spectra in Fig. 2.17. Thereafter the choice of heteroatoms in α-position (C/N/O/F) effects a strong alternation in the energy values of the original "*s*"-orbital in benzene as standard [68,88–92].

We were interested in the differentiating influence of "ACC"-substituents provoked by the type of heteroatoms in α-(C/N) or β-position (N/O/F), because the variations of such "ACC"-substituents alone render accessible a drug with the inverse effect of Nifedepine (vide Scheme 2.7) [93]. We investigated the consequences of "ACC"-substituents in the *para*-positions in diphenyl- and diphenylmethane derivatives both by ^{13}C-NMR-data and by regioselectivities of nitration in the *ortho*-position of the less substituted aromatic ring (see Fig. 2.18).

Starting point was the observation, that mono-nitration of diphenyl and diphenylmethane leads, under coherent conditions, to $\approx$ 2:1- respectively 1:2-*ortho-/para*-nitro-products (vide (a)). In (b) the ^{13}C-NMR-data of the *ortho*-C-atoms in the less substituted ring show little differentiated values at all, but equal patterns. The patterns of *ortho*-nitro-selectivities are inverse in contrast for both investigated systems. This means, that a new quality in the process has to be taken into account in one case [35]. One could say perhaps, in one example the patterns are coupled together analogously and in the second case inversely.

2.7 Conformational Changes in Phenyl Systems

Our proposed principle of investigation, of limiting the number of possible conformations by systematic variations in the subsystems – here following the direct unifying principle – and thereby controlling a definite influence e.g. on positions of equilibria and even reactivities by fixed arrangements avoiding compensations, is demonstrated only schematically [94].

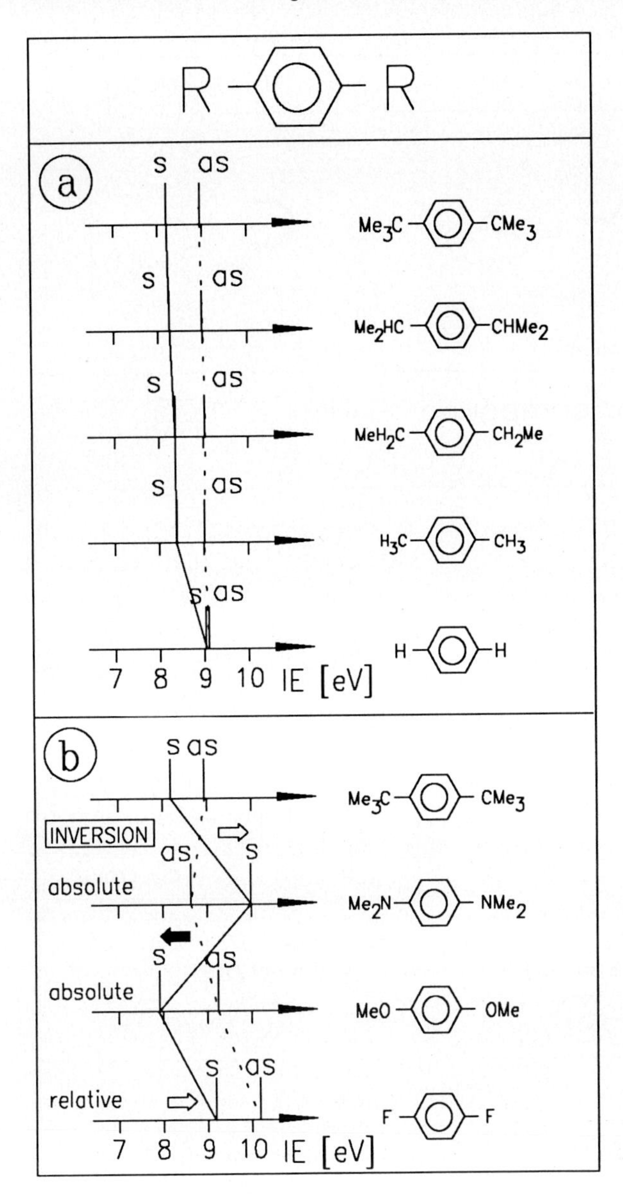

Fig. 2.17a,b. Energy levels of π-states in 1,4-disubstituted arenes [88,89]. **(a)** For alkyl substituents with increasing branching. **(b)** Values of π_s- and π_{as}-states as obtained from PE-spectra of considered 1,4-disubstituted arenes having heteroatoms from the second period of PSE as central atoms of these substituents [90,92]

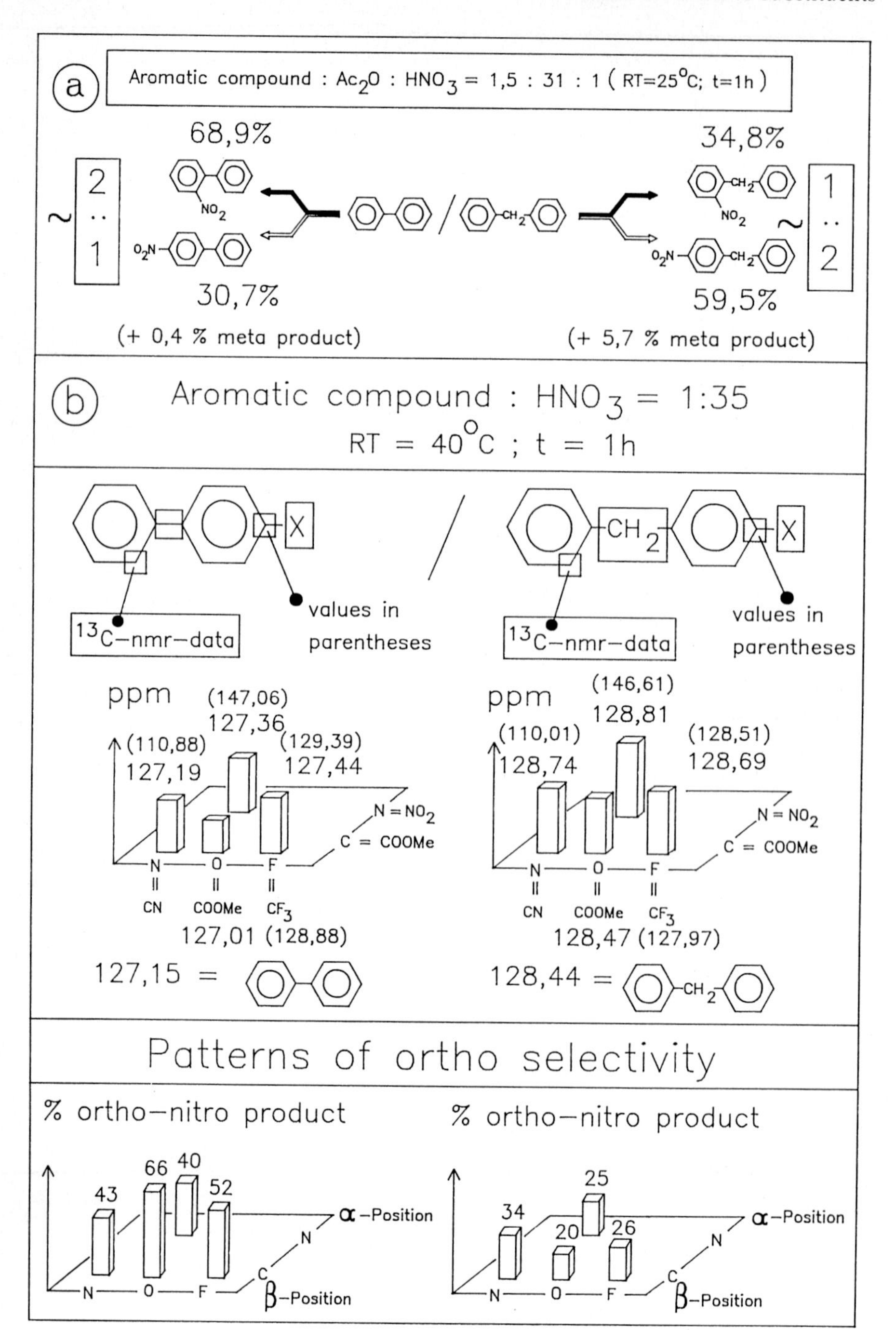
a
Aromatic compound : Ac2O : HNO3 = 1,5 : 31 : 1 (RT=25°C; t=1h)
68,9%
34,8%
~ 2 : 1
NO2
O2N
30,7%
(+ 0,4 % meta product)
CH2
1 : 2
59,5%
(+ 5,7 % meta product)
b
Aromatic compound : HNO3 = 1:35
RT = 40°C ; t = 1h
X
CH2
13C-nmr-data
values in parentheses
ppm
(147,06)
127,36
(110,88)
127,19
(129,39)
127,44
N = NO2
C = COOMe
CN
COOMe
CF3
127,01 (128,88)
127,15 =
(146,61)
128,81
(110,01)
128,74
(128,51)
128,69
128,47 (127,97)
128,44 =
Patterns of ortho selectivity
% ortho-nitro product
43
66
40
52
α-Position
β-Position
34
25
20
26

INHIBITION / ACTIVATION OF Ca^{++}-FLOW COUPLED WITH BLOOD PRESSURE

NO_2 / CF_3

COOR / NO_2

ROOC H CH_3 N H CH_3

Scheme 2.7

Such different conformations caused by type of perturbation as shown in the examples of Fig. 2.19 could easily be identified by X-ray investigations. We have depicted in (a) the change of conformations in -OMe/-N(i-Pr)$_2$ and phenyl groups depending on O/S at phosphorus atom [69,95]. In (c) is a quite general representation of possible con- and disrotatory movements of two phenyl groups, which are coupled to one center (or even to an axis!), starting from a propeller conformation. For this, three examples are selected in Fig. 2.19b. In the first example two different molecules in the elementary cell differentiate by con- or disrotation giving up degeneracy (by winning energy in the collective) [96]. In the second example, it is demonstrated by one and the same ligand in alternatively *cis-/trans*-complexes [97,98], that in every case other conformations are most cooperative. In the third example a degeneracy in a cluster formation is omitted in order to ensure the gain of energy in the crystal. The giving up of degeneracy in case three by choosing in each case two alternative conformations is demonstrated clearly in addition by the schematic representations of these two inverse conformations of the two subsystems n-butyl and the con-/disrotatory twisting of both phenyl groups in the two separated halves of the cluster [99].

Fig. 2.18a. Mononitration of diphenyl and diphenylmethane under coherent conditions. **(b)** The extreme small deviations among the δ^{13}C-NMR-data of the C-atoms in the ortho-position of the less substituted rings of the derivatives from the standard components diphenyl and diphenylmethane are represented – measured under strongly coherent conditions – in the form of patterns as function of the type of heteroatoms in β- (-C ≡ N, -COOR, -CF_3) and α-position (-NO_2) of the substituents. A comparison with the patterns of *ortho*-selectivity in the coherent mononitration (redox processes excluded) of both systems is only in accordance to the derivatives of diphenylmethane. Or to say it in other terms the patterns of ortho-selectivity are "symmetrically" or "anti-symmetrically" coupled with those of ^{13}C-data. The values for ipso-positions are given in parentheses

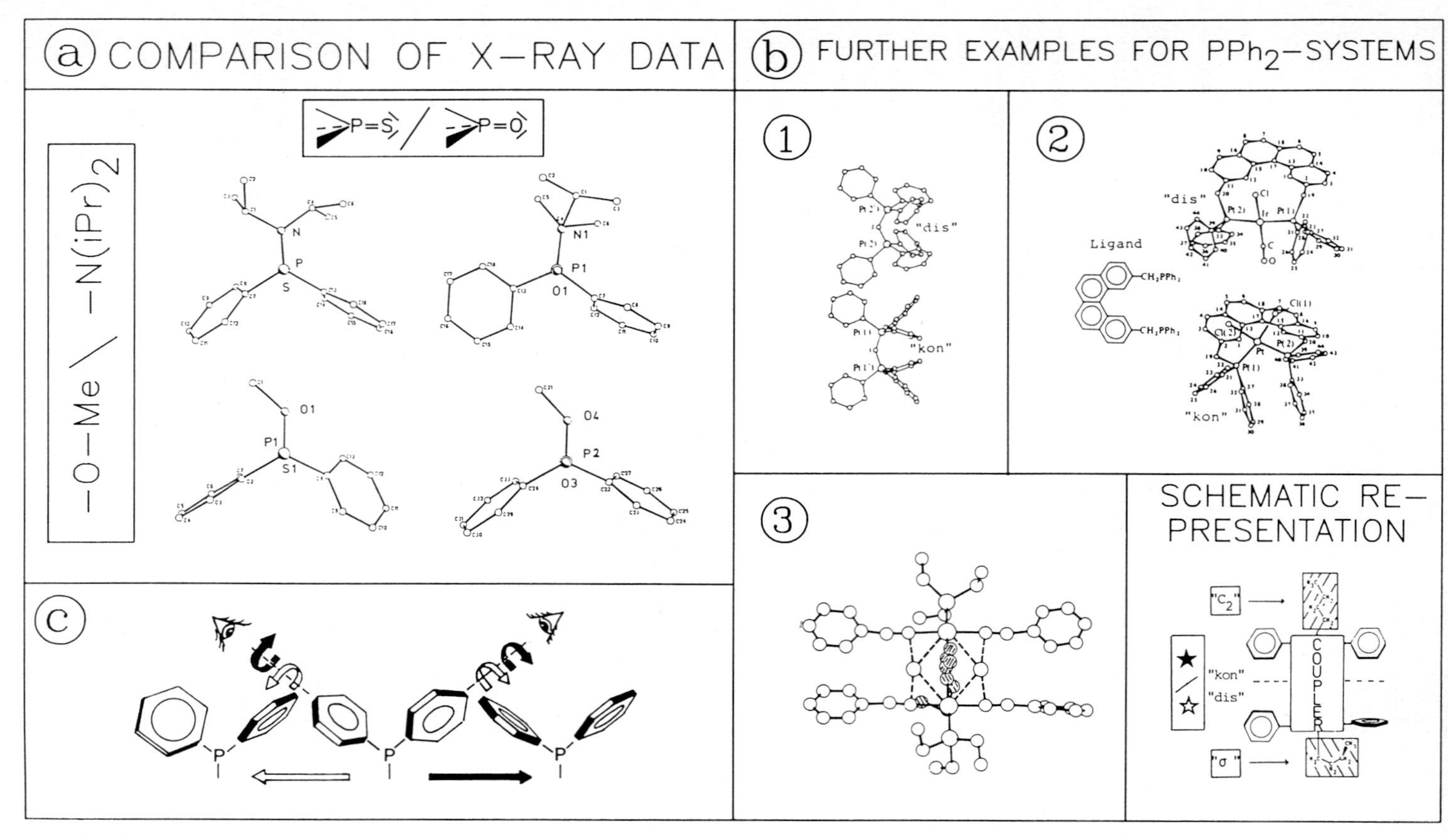

Fig. 2.19a. In every case inverse influence of P = O/P = S subsystems and –OMe/–N(i-Pr)$_2$ substituents on the conformations in the crystal [95]. **(b)** Three examples [96–98] for omitting degeneracy by alternative variations of conformations, which are in **(c)** generally represented by an example of two phenyl groups at a phosphorus atom

3 Examples of Absolute, Alternative Orders in Chemical Systems by Pairs and Alternating Classes of ALTERNATIVE PRINCIPLES

Die Natur scheint tatsächlich Sprünge zu machen – und sehr außerordentliche dazu.

Max Planck (1914)

We would like to introduce some ALTERNATIVE PRINCIPLES being exchanged in subsystems of educts, effectors, solvents etc. and thereby creating inverse orders in the considered states or processes in an unsteady way.

3.1 Examples for the Absolute, Alternative Effects by ACC/DO Heteroatoms

The central position of carbon in the *p*eriodic *s*ystem of *e*lements (PSE) corresponds to its predominant importance in biochemical systems and as standard relation in organic chemistry. A relatively greater or smaller first ionization potential of heteroatoms against carbon as standard is depicted in Fig. 3.1 as an important measure for their behavior [100]. All heteroatoms of stronger attractive power for electrons, which belong to the highest occupied orbitals, are classified as acceptors (ACC) and those with less attractive power as donors (DO). In choosing ACC/DO definitions one has to remember carefully in every case the reference (here free electrons) and the standard (here C).

The pairwise arranged heteroatoms N/P, O/S and X (Cl and Br)/I show always a distinct absolute ACC/DO behavior related to this standard carbon. Therefore it is obvious, that the hydrocarbons (CH), perturbated by O, N and X, very often obey the same rules and such ones, being perturbated by P, S and I, the inverse rules. The familiar reactions of organic chemistry with the well known rules of reactivity were jokingly called CHONX-chemistry by E. Negishi at the summer school at Swansea (Wales) in 1979. We, in contrast to this abstract term of CHONX-chemistry, of which the rules – very often one-sided – are offered at the beginning to undergraduates, would like to put forward the alternative term of CHIPS-chemistry to draw the readers attention quite precisely to the expected inverse rules (shaping coupled information from alternatives!). Later on we will slowly familiarize the reader with the importance of *relative* ACC/DO behavior to carbon by the paired atoms C/Si or N/O, too. Therefore all cases – though more often observed – are queued up in which a change of ACC/DO heteroatoms causes only relatively inverse effects in the behavior of chemical systems. For "becoming acquainted and familiarized" [47] we have chosen only those ex-

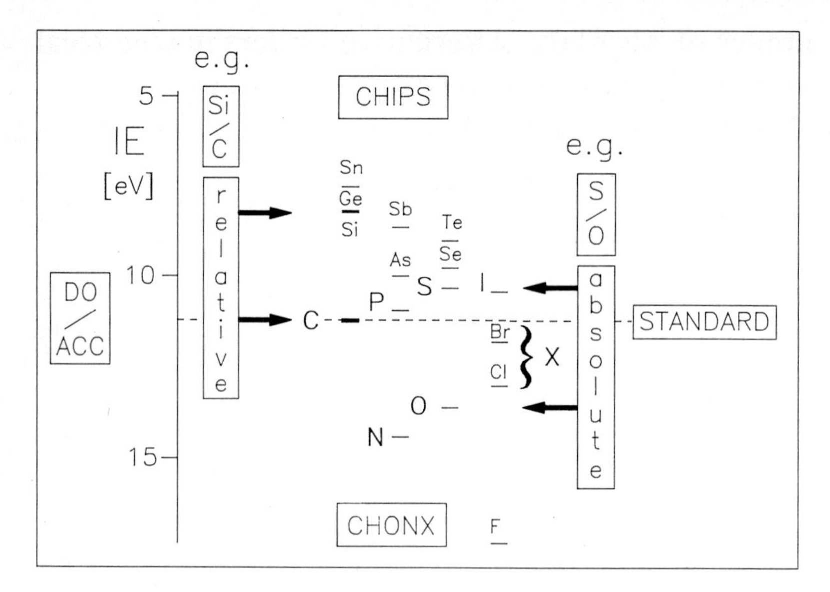

Fig. 3.1. The first ionization potential of carbon [100] as standard for the classification of ACC/DO heteroatoms (for definition CHONX/CHIPS vide Fig. 3.4 and text)

amples, in which a change of ACC/DO heteroatoms in fixed positions of a system leads to absolute inverse (alternative) effects in the structure or process.

Four examples of the effect of an O/S ATOM REPLACEMENT are presented in Fig. 3.2a: Li-*n*-butyl converts anisole to the *ortho*-lithiated arene derivative, whereas – according to [101] – the corresponding thio-compound forms a lithium-alkyl component (nucleus/side chain $\hat{=}$ π-/σ-reactivity). (b) Acetic acid ester is transformed to a Knoevenagel and the corresponding "carbanion" of the thio-ester [102] to a Michael product with cyclohexenone, whereby in the last case a C-S-bond is formed at first under kinetic control and a C-C bond under thermodynamic control [102]. (c) An inversion of regioselectivity is observed in the coupling of an alkoxi- or thioalkoxi-acetylene with a Cu-Li-vinyl reagent [103]. (d) X-ray analyses [32,95,104,105] demonstrate as alternatives for the dimers of allyl-Ni-X-complexes with X = alkoxi or X = thioalkoxi groups in bridgehead position – four-membered rings consisting of two metal atoms and the two heteroatoms O/S – a planar or distorted arrangement. In addition the two allylic fragments show a type of "chair" arrangement in the planar four-membered ring (O-system) and a kind of "boat" arrangement in the tilted one (S-system). We will discuss the influence of variations in the alkyl substituents at the heteroatoms in bridgehead position on the stability of both arrangements of the four-membered rings later on in context with Fig. 5.11a [32].

Further examples are shown in Fig. 3.3 for three, in every case, absolutely alternative structures caused by the choice of N/P or Br/I heteroatoms and for the

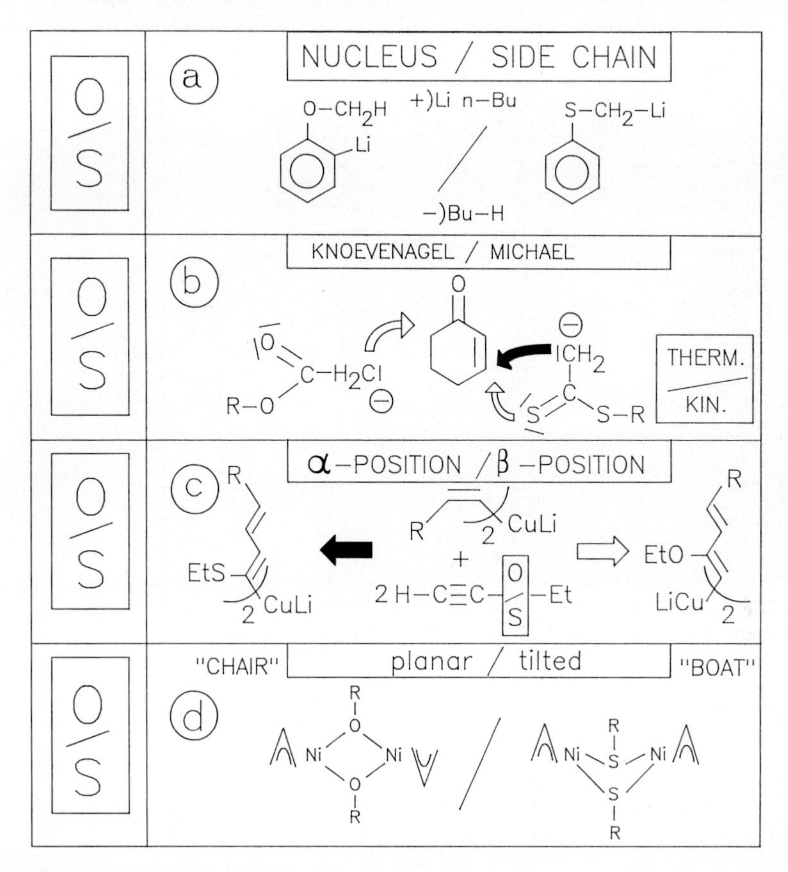

Fig. 3.2a-d. Four examples with absolute control of inverse order in processes depicted in **(a)** to **(c)** and in a structure in **(d)** by the choice of the heteroatoms O/S in equivalent positions of chemical systems

influence of Cl,Br/I on the stereochemistry of products in a polymerization process of butadiene at a H-X-modified Ni-catalyst [106]. In the case (a) we have to consider the fact, that beside the central atoms N/P the substituents (2 × H/ 2 × *t*-Bu) are changed, too. Finally for the collective crystalline system both the alternative M- and U-conformations are identified without doubt in the bis-keto- and keto-enole-tautomers. The trend is unequivocal in a series of spectroscopic investigations in solution, that the choice of heteroatoms N [107,108]/P [109,110] in the system causes the determining alternative effect. The arrangement LINEAR/TILTED ≙ 180°/90° (END ON/SIDE ON) is proved to be correct for the two examples in (b) [111] and [112]. Even in the dimminiation in (c) it is clearly discernible, that the 1,5-dihalo-chinones are "left/right" distorted in the crystal depending on the type of halogen atoms (Br/I) – as

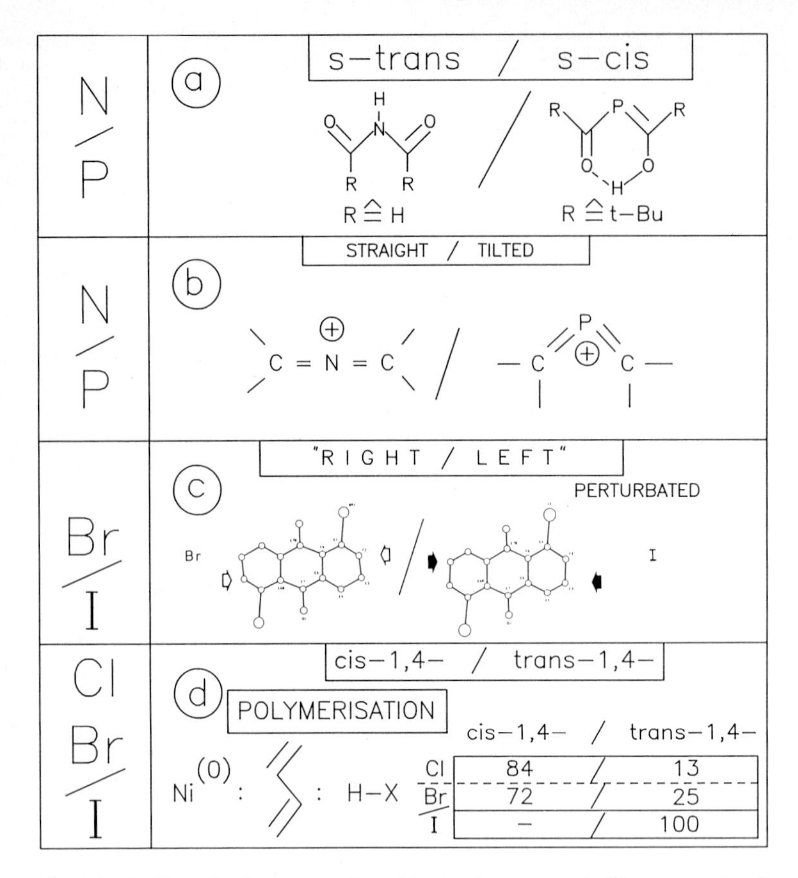

Fig. 3.3a-d. Four further examples with absolute control of inverse orders in structures in **(a)** to **(c)** and a polymerization in **(d)** by the choice of the heteroatoms of sectors AII/AIV (vide Fig. 3.4) such as N/P or Cl, Br/I

demonstrated by white/black arrows – [113]. (d) An absolute change in the stereoselectivity of the overwhelming 1,4-polymerization (*trans-/cis-*) of butadiene is observed by passing from hydrogenchloride and -bromide to hydrogeniodide as coeffector in Ni-catalysis [106].

3.2 Separation of Main Group and Transition Metal Elements in Four Sectors (PSE-sectors)

With respect to present rules of a thumb we have proposed [21], dividing the main group and the transition metal elements into four sectors (vide Fig. 3.4a and b). The separation into sectors with the standard carbon in the upper center of the

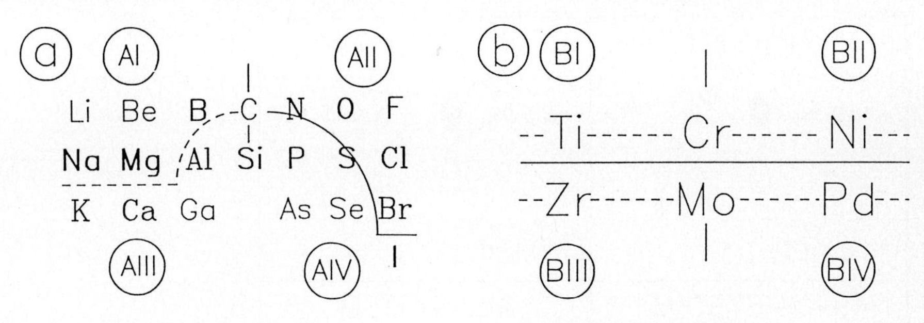

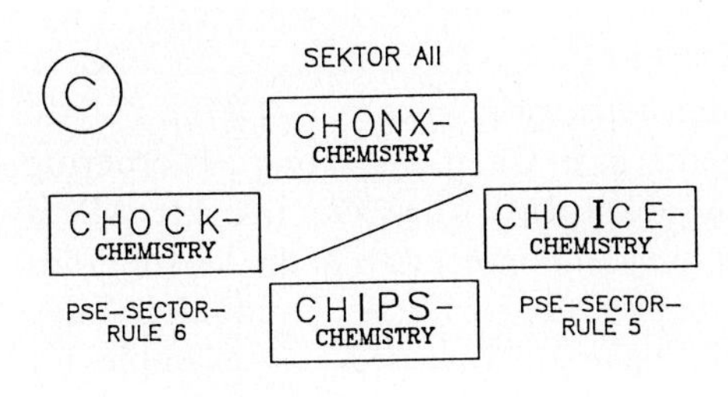

Fig. 3.4a-d. Separation of main group **(a)** and transition metal **(b)** elements in four sectors by ACC/DO and S/AS qualities (see Figures 2.7). **(c)** For definitions of the abbreviations CHONX/ CHIPS or CHOCK/CHOICE see context and PSE-sector rules in **(d)**. Rules of ATOM REPLACEMENT for the practicing chemist at fixed positions of a system (see Fig. 3.9, too)

(d)

PSE-sector-rule 1: Substitution of an atom near the boundary-line of a sector by an atom situated at a greater distance from it, may induce, in general, a change from covalent to ionic behavior.

PSE-sector-rule 2: The ATOM REPLACEMENT located near a common boundary-line of a sector in the Periodic System of Elements (PSE) leaves uninfluenced the kind of main interaction (orbital control). If both atoms are derived from the same sector of the Periodic System of Elements, the replacement normally leaves unaltered also the order of the system.

PSE-sector-rule 3: If ATOM REPLACEMENT is effected between atoms of two neighbouring sectors (near the boundary-line of the sectors), we have the highest chance to expect a change of the molecular order in structure as well as of the respective reaction pattern.

PSE-sector-rule 4: It is possible to compensate the effects of an individual ATOM REPLACEMENT according to rule 3 on a given chemical system of more constituents. In order to achieve compensation, a second ATOM REPLACEMENT in another suitable subsystem may be sufficient (intra- or intermolecular compensation of the effect of an individual ATOM REPLACEMENT).

PSE-sector-rule 5: A multifold ATOM REPLACEMENT in a complex chemical system according to PSE-sector-rule 3 can be also compensated intramolecularly. The introduction of a second series of heteroatoms into the same molecular system can approximately reverse the "break of symmetry" induced by the first series of heteroatoms and practically reestablish an originally more or less sensitive energetically degenerated state (CHOICE-Chemistry).

PSE-sector-rule 6: A multiple ATOM REPLACEMENT will not necessary lead to an increase of reactivity. On the contrary - it can be used also to block catalyzed reactions (CHOCK-Chemistry).

X: AI or AIII	IXI : AII or AIV
Ψ_2 — AS ≙ Ψ_1 ⫲ S	Ψ_2 ⫲ AS ≙ Ψ_1 ⫲ S
empty	filled
4q+2−SYSTEM	4q−SYSTEM

Scheme 3.1

main group elements is realized by the alternative principles ACC/DO and S/AS, which we recognized with the help of patterns in Chap. 2 as important ordering principles for the rough classification of substituents. The sector line AII/AIV is derived without difficulty from the first ionization potentials of the heteroatoms at the right hand side of carbon in the PSE, presented in Fig. 3.1, by the alternative principles ACC/DO. In Figs. 3.2 and 3 we have still discussed some examples for the importance of an ATOM REPLACEMENT referring to PSE-rule 3 (vide (d)) in the form of absolute inverse effects. The separation into the sectors AI/AII respectively AIII/AIV results without compulsion, because the heteroatoms at the right hand side of the – in old fashion termed [114] – fourth main group can contribute in addition e.g. a free electron pair and a center an α- or ω-position where, in contrast, those on the left have no free electron pair but only an empty center. The ordering principles in π-systems – enlarged by such alternative heteroatoms – change in every case both the number of disposed centers (EVEN/ODD or ODD/EVEN) and the symmetry of the frontier orbital (S/AS or AS/S) only by heteroatoms of the sectors AII and AIV (vide Scheme 3.1).

The slogans CHONX- and CHIPS-chemistry presented in Fig. 3.4c we have still interpreted some pages earlier in their translatable and abstract meaning. The slogan CHOICE-chemistry will be discussed as an example for PSE-sector rule 6 (by applying other pairs of alternative principles, too) in Chap. 8. Here we will demonstrate, that by choice, inhibition phenomena can be overcome by compensations and even patents may be skillfully circumvented. Naturally the CONCEPT OF ALTERNATIVE PRINCIPLES – and ATOM REPLACEMENT following PSE-sector rule 6 in only a partial aspect – may also be used to stop undesired catalytic decomposition e.g. of components for heat transportation in pumps for the flow of heat that means using a chock (CHOCK-chemistry) for this stopping. An example of PSE-sector rule 1 can be found in Fig. 3.22.

We proposed the borderline for the sectors AI/AIII empirically, because the pairs Na/K and Ca/Mg effect at both sides of biomembranes inverse order by their amounts [115] (vide Fig. 3.5a). In addition an especially big gap in the polarizabilities of atoms was observed by Kutzelnigg [116] at the borderlines proposed by us. Absolute inverse effects in structures (Figs. 3.5a and b) and

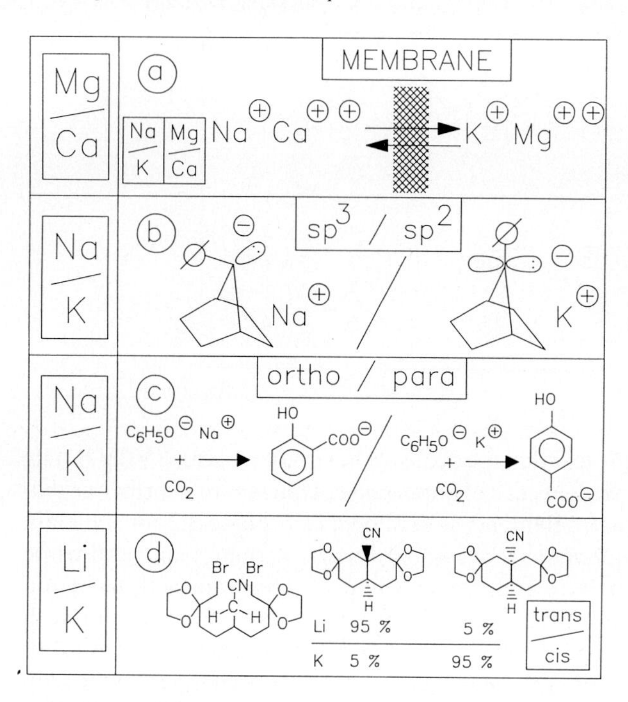

Fig. 3.5a. Starting from the fact, that low/high or high/low concentrations of Na^+/K^+ respectively Ca^{++}/Mg^{++} are to be observed inside/outside of cells at membranes [115] formed by building stones of one parity (vide Fig. 9.4), is presented in **(b)** an inverse structure of a "carbanaion" only by the choice of Na^+/K^+ counterions [117], in **(c)** an example for alternative regioselectivity (*ortho*/*para* by Na^+/K^+) in the Kolbe-Schmitt synthesis [118] and in **(d)** an example for the alternative regioselectivity of a twofold ring closure condensation (*trans-/cis*-decaline derivative by Li^+/K^+) [119]

processes ((c) and (d)) are induced only by the choice of alkali metal counterions. Whether a metalorganic compound with sp^2- or sp^3-hybridized carbon is formed by splitting off a methoxy group, depends only on the counterion Na^+/K^+ [117]. The supply of sodium or potassium phenolate with CO_2 determines in the Kolbe-Schmitt synthesis, whether *ortho*- or *para*-phenol-carboxylic acids are formed. An inverse control of *ortho-/para*-selectivity is caused by monosubstituted phenoles [118]; for this see also Scheme 4 in [29].

In (d) the cascades of two condensation reactions are determined in their stereochemistry of *cis-/trans*-decaline derivatives by the counterions Li^+/K^+ [119]. Similar stereodifferentiations are caused by Li^+/Na^+ counterions [120].

The separation into sectors, as illustrated by the examples with absolute effects in Figs. 3.5 b and c, is extremely useful. Previously we offered still other arguments in favour that the separation into the sectors AI/AIII doesn't lead to errors in logical typing. Starting from the Fig. 2.14 we pointed out with the

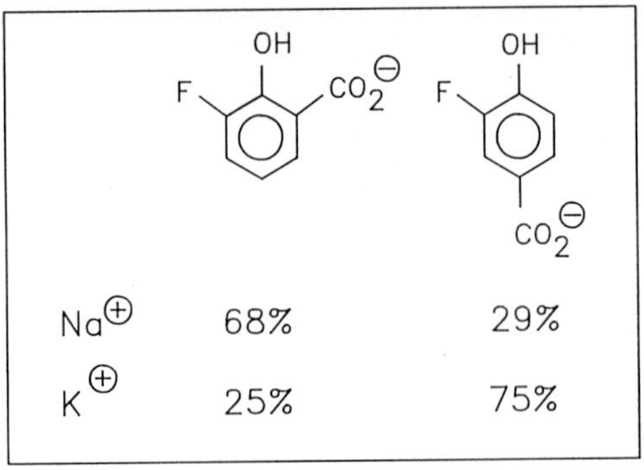

$Na^{\oplus}$	68%	29%
$K^{\oplus}$	25%	75%

following figures and by the text that the choice of heteroatoms with EVEN/ODD number of valence electrons is of crucial importance. That means in principle that the whole periodic system of elements – as shown in Scheme 3.2 for the main groups – has to be considered as a chessboard with alternate black and white squares with respect to phase relations (vide Chap. 5). This train of thought and the following have to be interpreted as closer explanations for PSE-sector rule 2. While from the left to the right hand side the number of valence electrons result in alternating phenomena, EVEN/ODD quantum numbers change from the top to the bottom. Here alternation phenomena have to be considered, too (vide Fig. 3.6). The alternation of enthalpies of formation for alkali-halides with a minimum stability of the sodium compound in Fig. 3.6a [121] corresponds to an inverse alternation in the reactivity of alkali metals (Li $<$ Na $>$ K) with a maximum at sodium in the Dieckmann condensation of e.g. different adipic acid esters under coherent conditions in (b) [122].

The sectors BI to BIV were separated empirically. As motivation for the borderlines BI/BII and BIII/BIV we offer the so-called "Lochformalismus" [123], whereby e.g. for d^8/d^2 complexes and so on in every case an inverse behavior has to be considered with respect to aspects of symmetry.

A change in phase relations by an ATOM REPLACEMENT BI/BIII and BII/BIV may be expected rather than in DO/ACC behavior of the transition metals relative to carbon. This inverse DO/ACC behavior has turned out rather to be true for the replacement of the metals Ni/Pd in π-allylic systems (vide Fig. 3.7). So Pd-[124,125] or Ni-catalysts [13,125] lead to a synthesis of alternatively

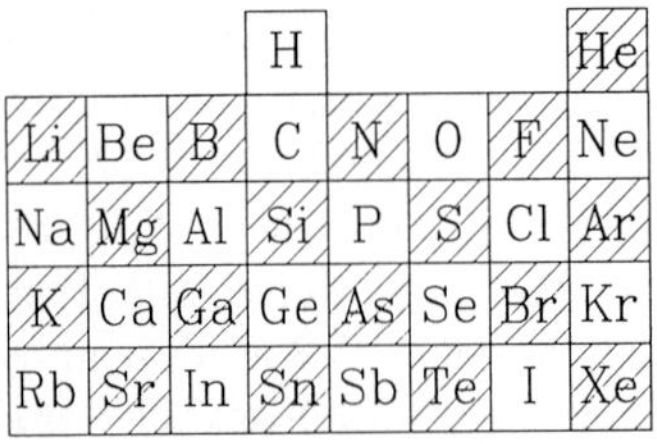

			H				He
Li	Be	B	C	N	O	F	Ne
Na	Mg	Al	Si	P	S	Cl	Ar
K	Ca	Ga	Ge	As	Se	Br	Kr
Rb	Sr	In	Sn	Sb	Te	I	Xe

Scheme 3.2

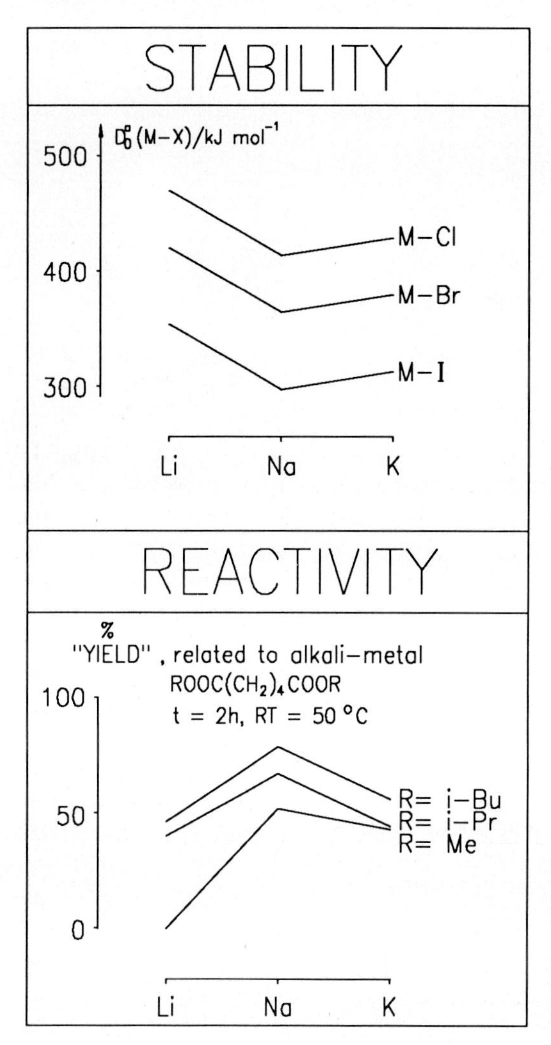

Fig. 3.6a. The enthalpies of formation of the alkali-halides show a striking, pattern [121]. **(b)** The maximal instability of sodium halides corresponds in a certain sense to the maximal reactivity of sodium-organic intermediates in dioxane under coherent conditions in the Dieckmann cyclization starting from different esters of adipic acid [122] ("yield = conversion)

n-octa-1,3,7-trienes or -1,3,6-trienes. Alternatively a "hydride" is transferred formally to a 3-position or a "proton" to a 1-position to an allylic intermediate – marked by lines – in (a). In (b) products are formed in a cross-dimerization of butadiene and 1-alkylsubstituted 1,3-dienes, in which an alternative C-C-coupling is observed to straight (Pd [125]) or branched (Ni [125]) codimers. Inverse patterns for the distribution of the methyl group in the C_8-chain occur in the co-dimeric methyl-n-octatrienes from butadiene and isoprene.

Alternatively CYCLIC/OPEN 2:1-codimers with different structures [125–127] are formed in (d) at Ni-/Pd-catalysts from two molecules of butadiene (forming a C_8-chain) and from oxo-compounds or Schiff-bases. The hetero-olefines react either in 3,6-positions of the marked α,ω-bis-allyl-chains under

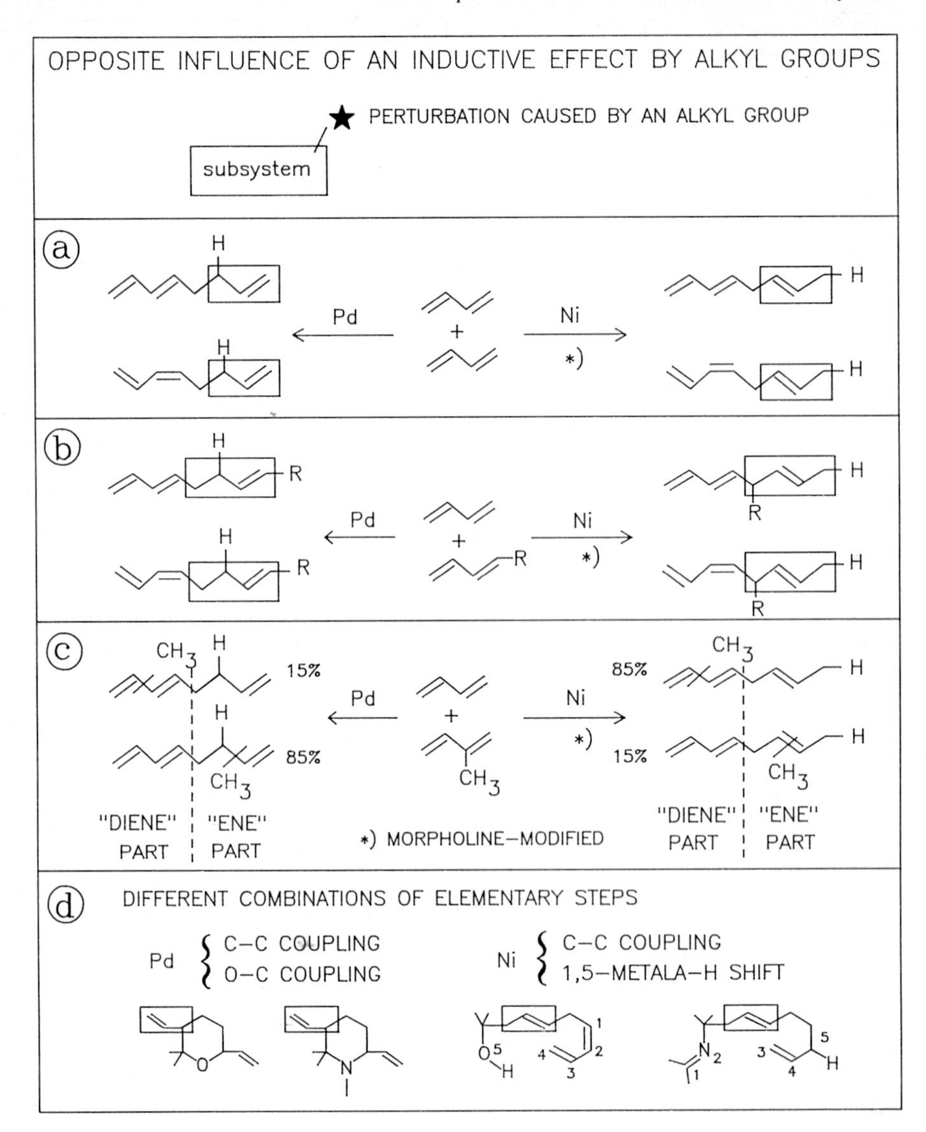

Fig. 3.7a-d. Four comparisons of alternative orders in the reaction products of Pd-/Ni-catalysts (borderline BII/BIV) in the synthesis of openchained dimers from two butadienes in **(a)**, from butadiene and octatriene in **(b)** respectively from isoprenes in **(c)** as well as in the 2:1-co-oligomerization of two butadienes with oxo-compounds or Schiff bases in **(d)**

cyclization or C-C-coupling in 1-position followed by a hydrogen-transfer reaction. Analogous reactions such as the cyclotrimerization of butadiene and the co-oligomerization of butadiene and but-2-yne are presented in Fig. 3.8 for the comparison of Ni-/Cr-Ti-catalysts. It is justified to discuss in an abstract manner HOMO-/SOMO-/LUMO-catalysts, which effect by the change in the type of activation step in every case an inverse order or – as in the borderline-case of Cr-catalysts – both possibilities of reaction are more or less of the same importance [128].

3.3 Errors of Logical Typing in DO/ACC Alternatives

Previously we demonstrated by example (a) in Fig. 1.1 the difficulty of transferring the term ACC/DO valid for atoms (vide Fig. 3.1) to classes of atoms, too, e.g. the substituents -COOR and -OR. The intrinsic difficulties of understanding we would like to elucidate with Fig. 3.9 with the help of the concept of ATOM REPLACEMENT. To identify the crucial ALTERNATIVE PRINCIPLES it is very important to introduce a σ-/π-separation and to define the corresponding carbon-π-system as standard. Starting from ethene and benzene π-systems result, which are characterized by the ALTERNATIVE PRINCIPLES EVEN/ODD number of carbon-centers and 4q-/4q + 2-π-electrons (EVEN/ODD number of π-electron pairs). In addition the subsystems – marked by lines – can be made alternatively electron deficient (electrophilic)/electron rich (nucleophilic) or vice versa by replacement of a carbon atom – here always in ω-position – by ACC or DO heteroatoms related to carbon as standard. In doing so we have to consider, that π-systems with EVEN or ODD centers differ in the HOMO/LUMO distance in a 2:1 ratio according to the simple HMO-theory [42]. The HOMO's are either bonding or non-bonding. The multifold combinations of ALTERNATIVES show impressively, that errors in logical typing can not be excluded by oversimplification.

3.4 The Alternative Principles EVEN/ODD

The importance of EVEN/ODD centers is not restricted to π-systems (normally one electron per center) but has an alternative meaning for σ-systems, too, proved by the four examples in Fig. 3.10. The well-known alternation in melting points of hydrocarbons – dependent on the number of methylene groups – is distinctly marked in the clearing points of liquid crystal phases [129] (example (a)). Overwhelmingly products of head-to-tail-/tail-to-head- respectively head-to-head/tail-to-tail-coupling are formed in the photochemical intramolecular dimerization of the double bonds in hexa-1,5-dienes or hepta-1,6-dienes following the "rule of five" [130] (vide (b)). Duplication of the bridges between the

a Cyclotrimerization of butadiene by means of Ni-, Cr- or Ti-catalysts

Substrate	Products	Catalyst Ni	Cr	Ti
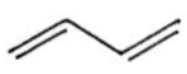 butadiene	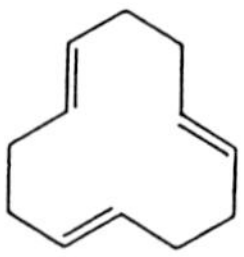 cyclododecatriene (ttt)	95%[a]	60%	5%
	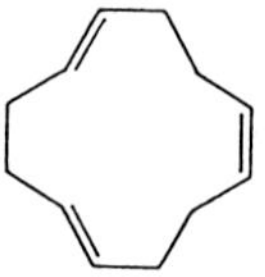 cyclododecatriene (ttc)	5%	40%	95%

[a] We added pyridine in a ratio pyridine:nickel = 1: 10 to avoid tcc- CDT formation

b Co-oligomerization of a 2:1 mixture of butadiene and 2-butyne at 40 °C by Ni-,Cr- and Ti-catalysts (the product yields (%) are based on alkyne)

Substrate	Products	Catalyst Ni–L	Cr	Ti
 butadiene + $H_3C–C{\equiv}C–CH_3$	2:1	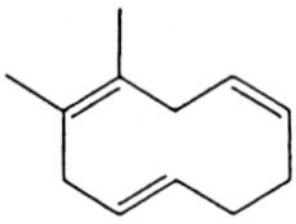 > 90%	30%	–
	1:1	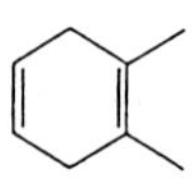 –	5%	45%
	1:2	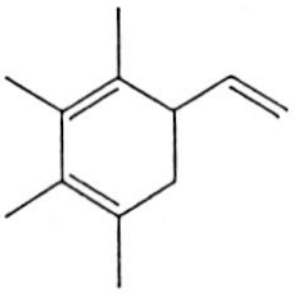 –	10%	30%

Fig. 3.8a-b. Ni-/Cr-/Ti-catalysts in comparison; **(a)** control of stereoselectivity in the catalytic cyclotrimerization as well as in **(b)** of the co-oligomerization of butadiene with but-2-yne. Ti and Ni belong to sector BI/BII ("Lochformalismus") with Cr as a borderline-case in between

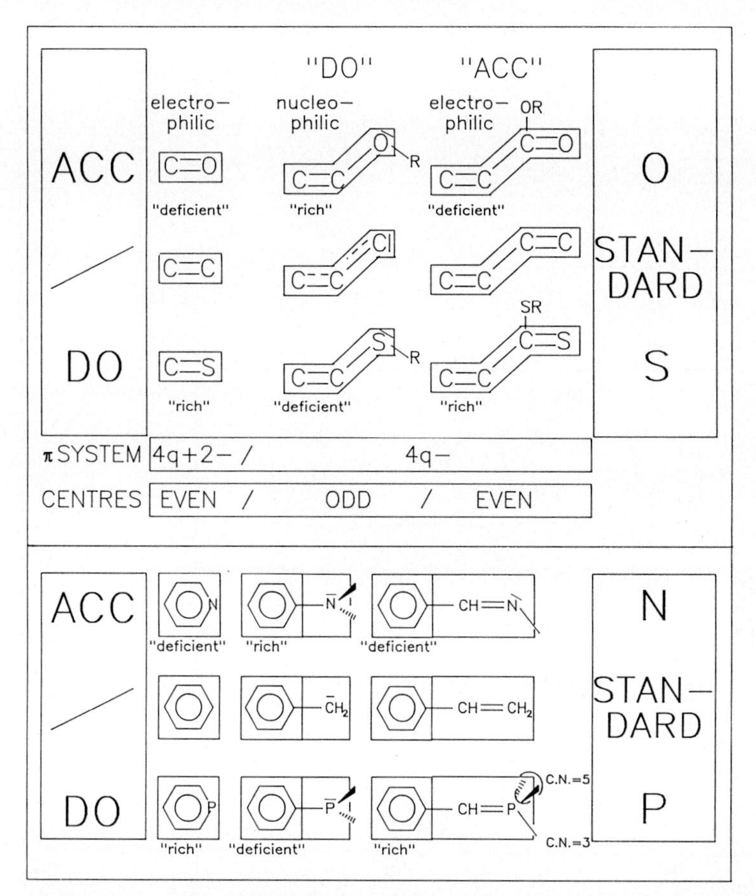

Fig. 3.9. Characterization of parent-systems with substituents as corresponding π-systems – based on the all-carbon standard system – with 4q–/4q + 2-π-electrons having EVEN/ODD centers and perturbated by ACC/DO heteroatoms in ω-positions. "DO/ACC" classifications of substituents such as $-OCH_3/-COOCH_3$ lead to errors in logical typing

double bonds provokes an absolute preference of one or the other type of photochemical four-membered ring synthesis in *cis, cis*-cycloocta-1,5-diene or *cis, cis*-cyclodeca-1,6-diene [131] (see top of the next page!).

In complex phospha-ferra-substituted enolates a threefold ethyl-/methyl-substitution at the phosphorus leads to alternative O/C-methylation [132] (vide (c)). We point out by (d), that a two-fold association of TPP at a $Ni^{(0)}$-catalyst (ML_2) induces a [4+2]- but a mono-association (ML) a [4+4]-cycloaddition of butadiene (see also Chap. 6).

In Fig. 3.11 we exposed a further general aspect of the alternatives EVEN/ODD. By the examples in Fig. 3.11 we have further drawn the attention to the fact, that it is better to speak of alternating homologous series than of homologous series. Often it is even better for consistent understanding to separate

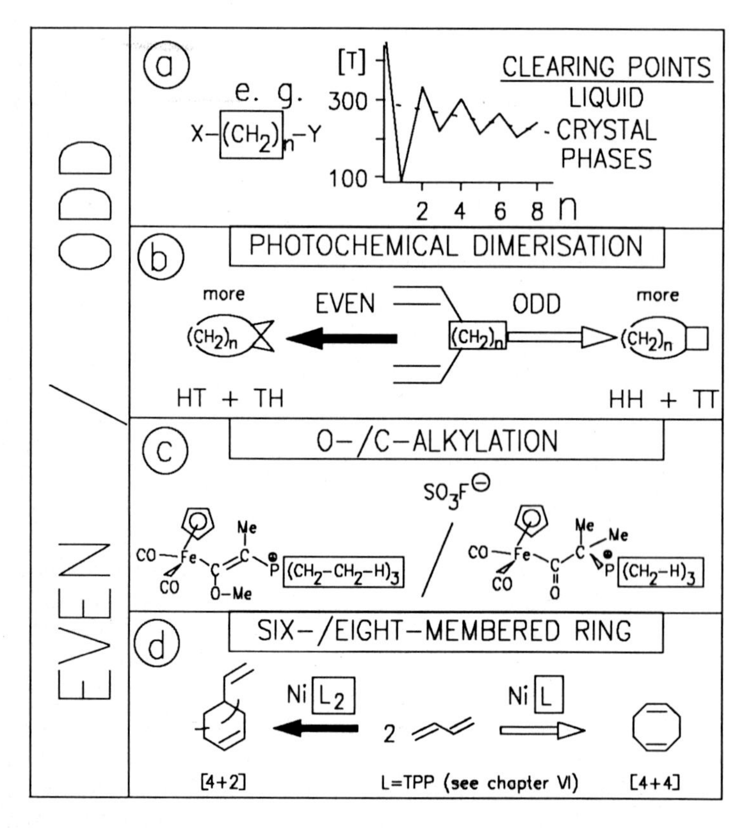

Fig. 3.10a-d. Four examples of alternations looking at clearing points **(a)** [129], in head-to-tail or head-to-head-coupling of photochemically activated olefines **(b)** [130], in an alternative regioselectivity **(c)** [132] and an eight-/six-membered ring-synthesis **(d)** by the choice of an EVEN/ODD number of methylene groups (**(a)** to **(c)**) or of ligands per metal in **(d)**

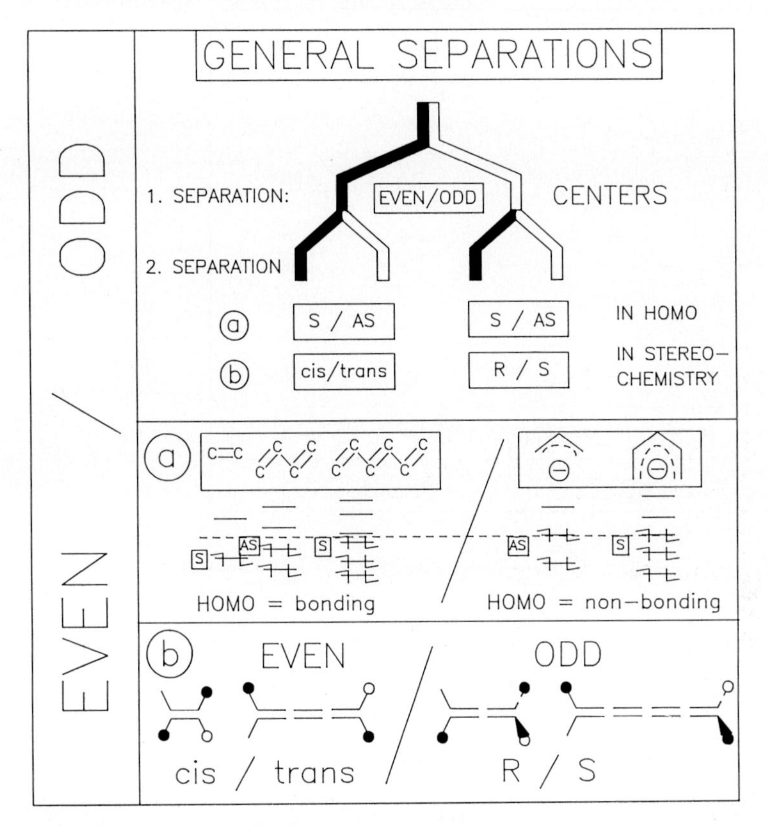

Fig. 3.11a,b. In the classification of substituent series it is demonstrated by two cases in **(a)** and **(b)**, that it is better to speak of alternating homologous series or even series of compounds with EVEN and ODD centers, whereby both series show different alternatives in stereo-chemistry, than to use the term homologues

the series into those with EVEN/ODD centers. We have pointed further to the importance of such separations by the examples (a) in Fig. 3.9. In Fig. 3.11b one recognizes that other stereochemical alternatives (*cis/trans*- or *R/S*) have to be considered in these separated systems depending on corresponding substitution.

Verhoeven [133] described the eminent importance of alternative conformations (see Fig. 8.3 and context) for α,ω-diradicals for the effect displayed by EVEN/ODD centers. The alternation of phase relations (S/AS/S/AS) in these diradicals is evident in Fig. 3.12a in their all-antiplanar conformations beside their relatively alternating stabilization and destabilization gained by model calculations. An absolute FORBIDDEN/ALLOWED arises in (b) e.g. for ring closure reactions in the alternative conformations of highest energy content.

As can be derived from Fig. 3.13, the symmetry rules are not restricted to formally pure π-systems [48]. But the reactions in these examples (a) to (d) require increasingly higher activation energies.

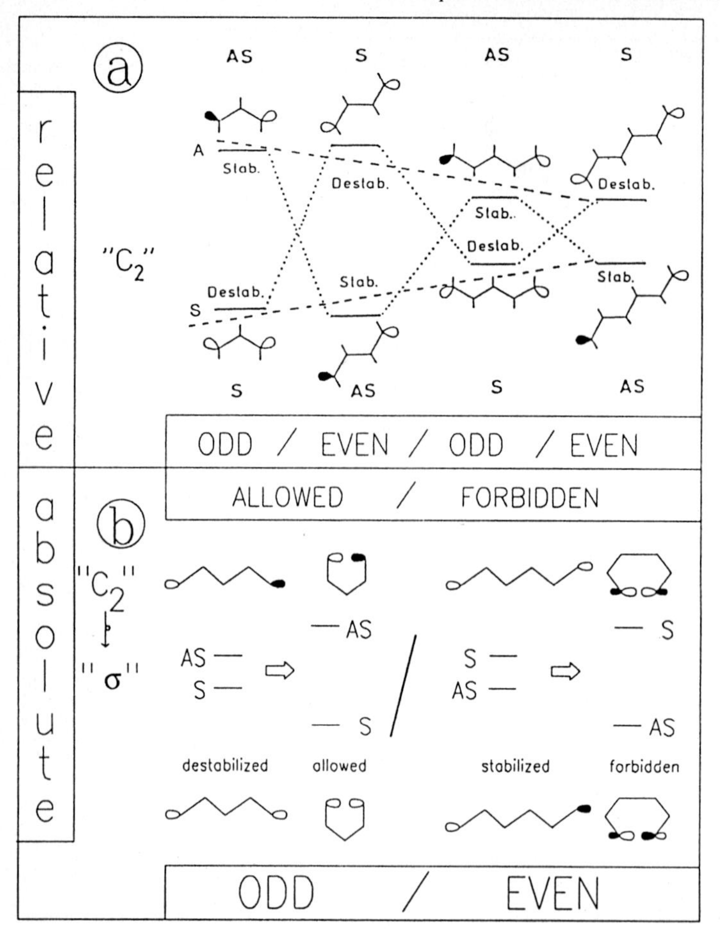

Fig. 3.12a. The phase relations in homologous α-, ω -diradicals show alternation in these positions and in addition relatively pronounced, alternative destabilization/stabilization. **(b)** By changing zig-zag arrangements (all-antiperiplanar) to "ring-forming" conformations (all-synperiplanar) the symmetry rules become absolute: forbidden/allowed (according to [133])

3.5 Aspects of Coupling Chemical Subsystems: The Alternative Principles OPEN/CYCLIC

With the direct unifying in perturbation theory [41] (see the individual steps in Fig. 3.14) one has a recipe for a simple procedure as a way of thinking to duplicate e.g. the number of centers in a π-system. This can be realized experimentally only by accepting long detours (pathways from two ethenes to butadiene).

In metalorganic chemistry π-systems may be coupled to larger units via the metal as coupler especially by easily verified associations. So ethene molecules in

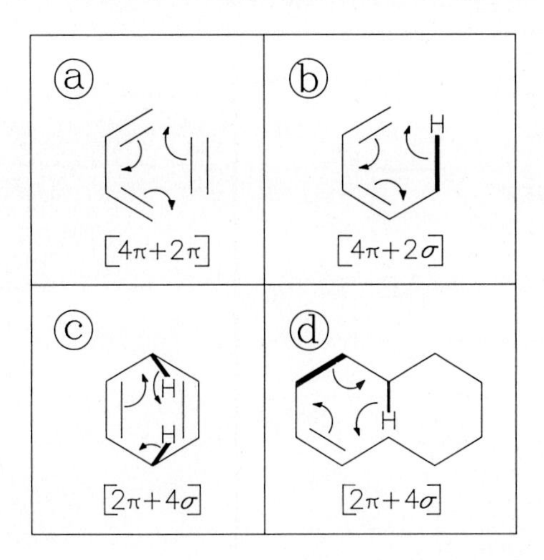

Fig. 3.13a-d. Analogous symmetry rules for systems without π-/σ-separation (see text, too) with increasing activation energy for the processes **(a)** to **(d)**

cis-position of Rh-complexes behave like butadiene in their coupled vibrations [134]. The difference of bis-π-allylic complexes compared with arenes is analyzed in detail [39].

The coupling of subsystems bears a general significance as an ordering principle, as we wish to demonstrate by the exemplified coupling of two ethene molecules by different couplers in Fig. 3.15. Nevertheless every coupled site is controlled by phase relations (the laws of general symmetry restrictions). Hereby the formal procedure of the direct unifying is always very important (vide Fig. 7.6, too) excluding the last case of metalalogy principle [39].

The direct unifying principle has to be imagined for recognizing the alternative principles OPEN/CYCLIC, too ([40], Scheme 12). Considering only the conformations in a bond of highest and lowest content of enthalpy and transfering these considerations to the two O-C-bonds in diethylether results in the conformations represented in Scheme 3.3. Diethylether at 20 °C exhibits around 90% of the all-*anti*-periplanar conformation A [135]. But the conformation C – despite a small folding in the five-membered ring – may be fixed by direct unifying. Thereby diethylether and THF represent two solvents with nearly inverse order of conformations; this should be of relevant importance for interactions controlled by symmetry. This is demonstrated in the examples (a) to (c) in Fig. 3.16 by comparison of the effects of the solvents Et_2O/THF. The inverse regioselectivity is determined in all three cases [136,137,138] by the choice of solvent. But in (c) the choice of the alkyl-groups (*n*-Bu/*t*-Bu) may have a higher hierarchical rank than the choice of the solvent. This is still an open problem. It is remarkable in case (d) of the electron rich olefine [139] that the PE-spectra of the olefines with OPEN and CYCLIC structure differ hardly [140]. While the olefine with OPEN structure forms a chromium-carbonyl-amine complex, a carbene-chromium-carbonyl complex is synthesized by C = C-splitting of the CYCLIC olefine. A P^I-compound may be inserted only into the double bond of an electron-rich

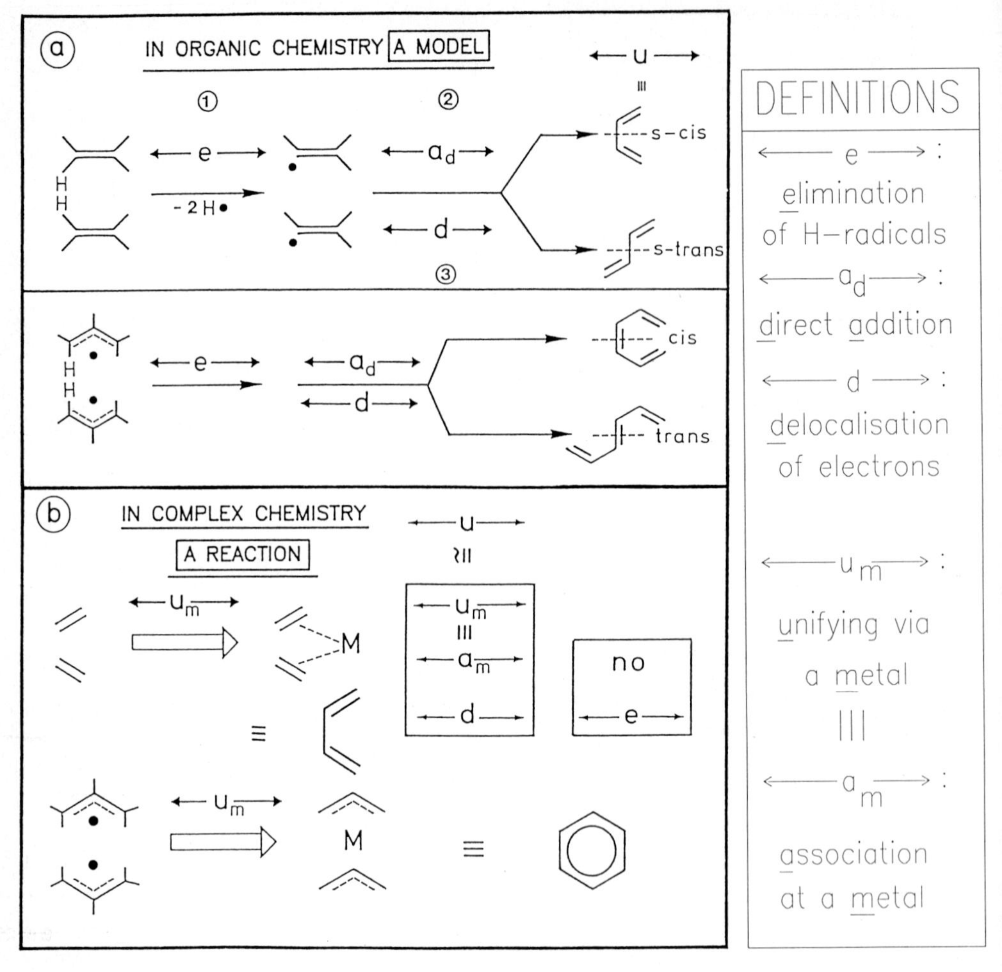

Fig. 3.14a,b. Detailed description of the stepwise procedure in the model of direct unifying in organic chemistry in **(a)** and in a real metal-analogous coupling of π-systems in **(b)**

−1 −1 A	+1 B
+1 C −1	D +1 +1

For this see also Fig. 10.1

direct unifying

Scheme 3.3

Type of unifying for ethenes			
Coupler	Examples	Symbol	Trivial name
direct		$\leftarrow U \rightarrow$	Unifying principle
CH_2–CH_2		$\leftarrow U_e \rightarrow$	Ethylogy principle
		$\leftarrow U_v \rightarrow$	Vinylogy principle
		$\leftarrow U_{o\text{-}ph} \rightarrow$	Phenylogy principle
		$\leftarrow U_{p\text{-}ph} \rightarrow$	Phenylogy principle
M	M	$\leftarrow U_m \rightarrow$	Metala–logy principle

Fig. 3.15. Different couplers for two ethenes applying the unifying principle and the metala-logy principle

olefine – as observed by Schmidpeter et al [141] – with CYCLIC structure. The olefine with OPEN dialkylamino substituents does not react in the same manner; nevertheless the desired compound is accessible by other reaction pathways [142].

Another important pair of solvents is 1,2-dimethoxi-ethane and dioxane (vide Fig. 5.1: OPEN/CYCLIC). Further investigations and examples for the alternative principles OPEN/CYCLIC are collected in Figs. 4.16a, 5.10d, 5.11, 6.12 and 13, 8.7 and 8.10 as well as 8.11.

3.6 The Metala-Logy Principle

According to Fukui [143] the metal may act as catalyst either in the function of a coupler for the substrates to be unified or in function of a perturbating subsystem, which changes the property of one of the two substrates in such a way, that an interaction and thereby a reaction is provoked.

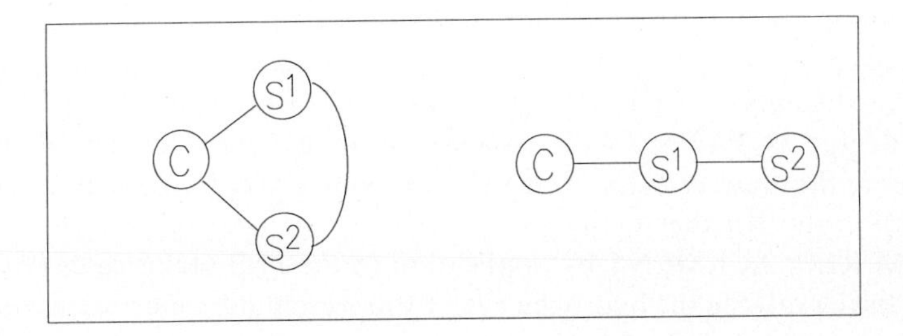

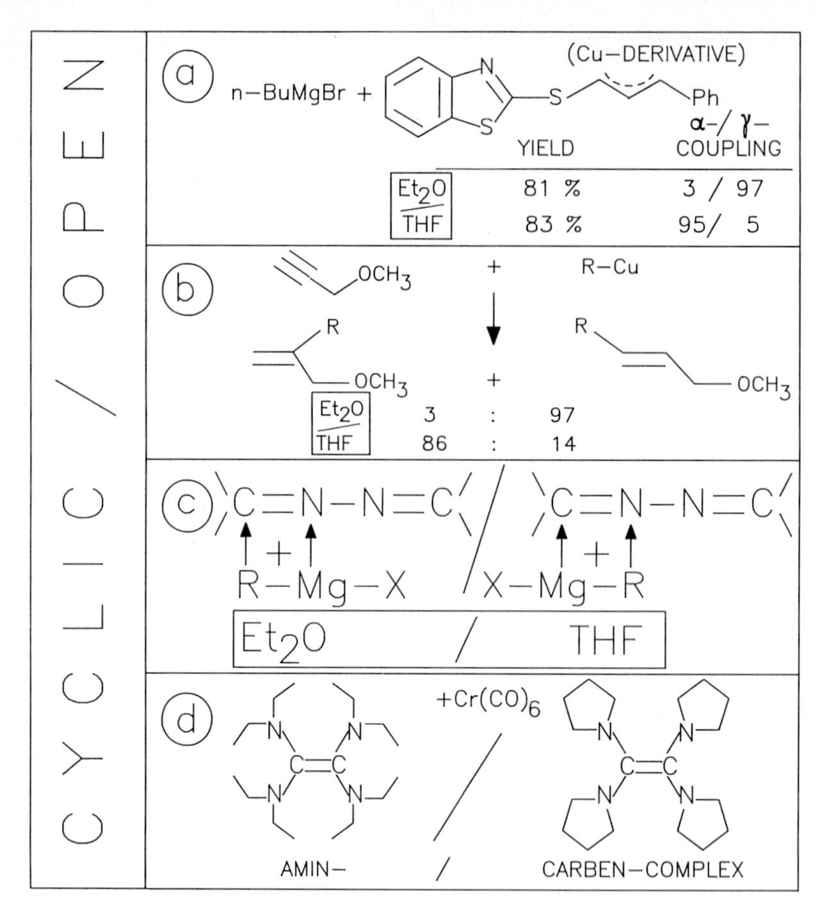

Fig. 3.16a-d. Four examples demonstrating the importance of CYCLIC/OPEN structures. The alternative choice of the solvents Et_2O/THF provokes in the cases **(a)** to **(c)** a change of regioselectivity in the products. **(d)** Electron rich olefines behave totally differently depending on structure

This idea of the metal as a coupler is presented in Fig. 3.17 prepared for a practicing chemist in such a way, that the catalyst and the two substrates to be unified are separated in two subsystems. Alternative variations may be verified *at* and *in* the subsystems. The variation at the metal as catalyst comprehends the alternative change of the ligand field in a far reaching sense by the properties of one or more ligands via ALTERNATIVE PRINCIPLES (vide e.g. Figs. 6.12 and 13), by degree of EVEN/ODD associations or generally by the alternative principles of the amount MUCH/A LITTLE (vide Chap. 6). The catalyst metals or the carbon atoms in the organic substrates are exchanged following the concept of ATOM REPLACEMENT by application of the PSE-sector rules (vide Fig. 3.4). We have replaced the hydrogen even by representative substituents (see e.g.

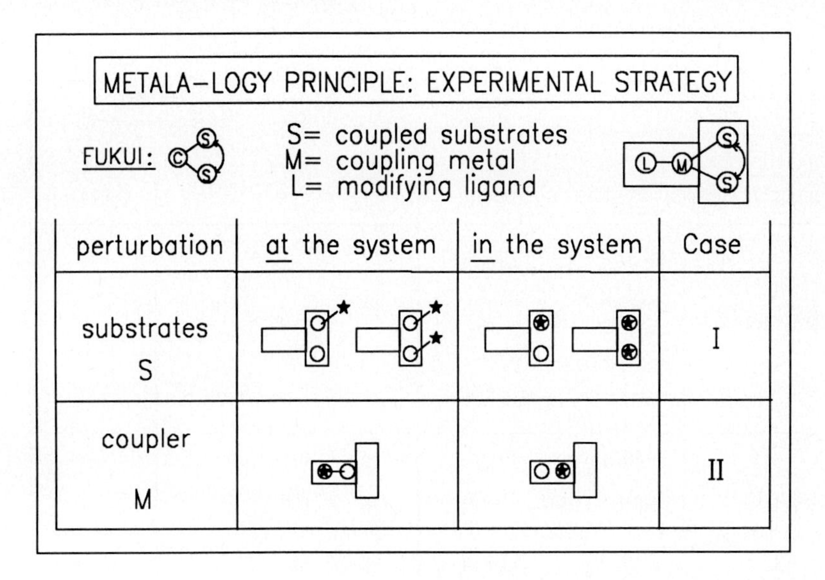

Fig. 3.17. The catalytic system is formally separated into metal + ligand and the substrates to be coupled together, to achieve in systematical variations *in* (ATOM REPLACEMENT) and *at* (REPLACEMENT of substituents and ligands) the system

Fig. 2.10d) and thereby extended the variations *at* the substrates shown as real examples in Fig. 3.18. A great variety of syntheses is described elsewhere [11,12] and only a selection is presented in Fig. 3.19. In (a) the oligomerizations and co-oligomerizations of olefines are shown at $Ni^{(0)}$- and $Ni^{(0)}$-L-catalyst [12]. In (b) some co-oligomerizations of butadiene and alkynes are presented. So butadiene may be coupled with but-2-yne by choice at a $Ni^{(0)}$-TPP-[144] or a Ti-catalyst [128] to a 2:1-adduct, which is transferred to a *cis*-4,5-divinylcyclohexene in a Cope-rearrangement, or a 1:2-adduct, 5-vinyl-cyclohexa-1,3-diene, which forms a tricyclo-octene [145] in an intramolecular Diels-Alder reaction. The synthesis of 1:2-adducts may also be performed at a $Ni^{(0)}$-TPP catalyst, when an electron-deficient acetylene-dicarboxylic acid diester is converted in the presence of butadiene instead of an electron-rich but-2-yne [12]. Application of cyclic mono- and di-acetylenes leads after reaction of two butadienes per triple bond e.g. to a ring enlargement of eight or sixteen carbon atoms [12,147] by a combination of Ni-catalyzed and classical reactions [146]. The codimerization of 2,3-dimethyl-buta-1,3-diene and but-2-yne is achieved with 1,4-diazadiene-Fe-catalysts [12,153] in high selectivity.

Both substrates have substituents which increase electron density at the π-system and a thermally initiated Diels-Alder reaction is made impossible. The catalyst takes over redox functions and makes possible this unusual

EXAMPLES FOR EXPLORATION OF CATALYTICAL SYSTEMS

★ perturbation	at the system	in the system	Case
substrates S	$-CH_3$; CH_3, COOR; CH_3; CH_3; H_3C-, $-CH_3$; $H_3C-\equiv-CH_3$ $H_3C-\equiv-COOR$	O; N–R; O, CH_3; N–R; N; N, N	I
coupler M	$P(OR)_3$ PR_3 HPR_2 HNR_2 HOR $Al(OR)_3$ $B(OR)_3$	Ni / Cr / Ti Ni / Pd Ni / Fe	II

Fig. 3.18. Realized variations in subsystems as proved in Fig. 3.17

Me, Me + Me–≡–Me —Fe(N,N)→ Me, Me, Me, Me

[4+2]-addition. These possibilities of catalyzed processes are not at all exhausted.

A few co-oligomerizations of butadiene with hetero-olefines are presented in Fig. 3.19c [127,148,149]. The decisive influence of methyl groups in 4- or 2-position of acroleine is by no means an exception in the $Ni^{(0)}$-L-catalyzed 2:1-co-oligomerization with two butadienes to products of Knoevenagel or Michael type (see results in Fig. 5.5). At this moment, we would like to discuss a little bit more generally the importance of perturbations in alternative positions of chemical systems.

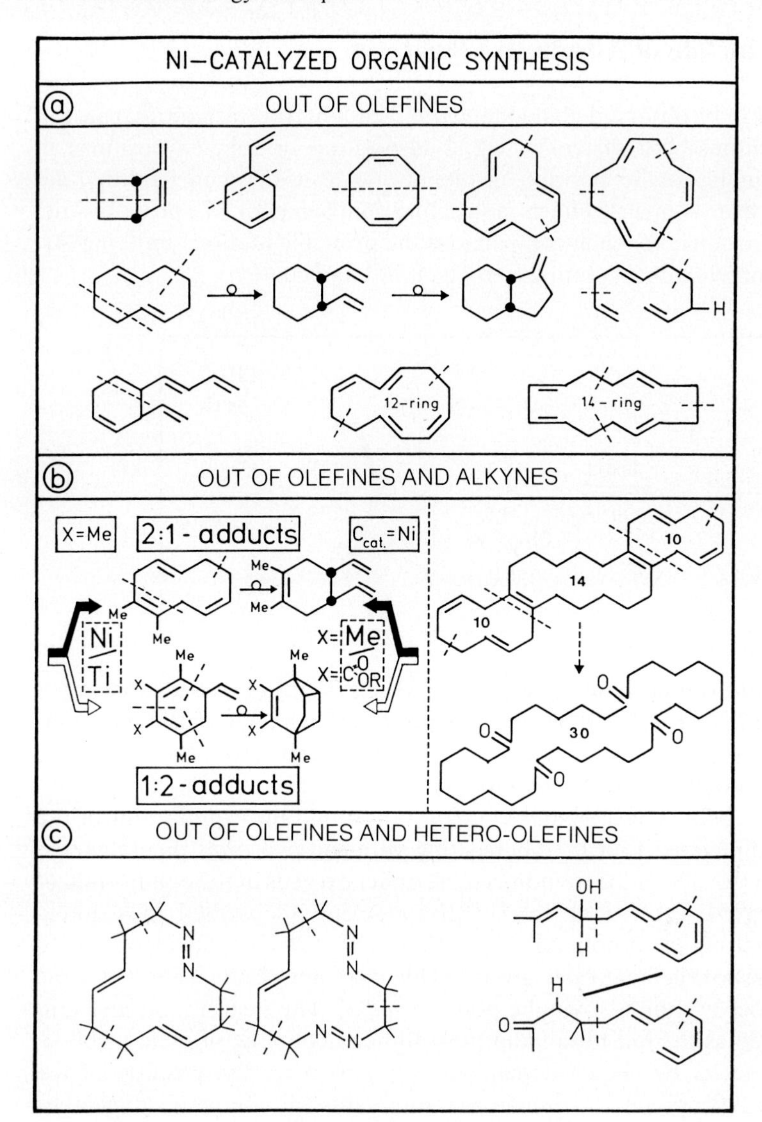

Fig. 3.19a-c. Organic syntheses of high selectivities at $Ni^{(0)}$-catalysts by oligo- and co-oligomerization of olefines in **(a)**, co-oligomerization of olefines and alkynes in **(b)** as well as olefines with hetero-olefines in **(c)**

3.7 The Principle of Alternative Positions

Very often it can be observed, that identical substituents (perturbations) cause in different positions of a system – e.g. 1-/2-position or 1,4-/2,3-positions in 1,3-dienes, vinylic/allylic position in olefines, α-/β-position in naphthyl-derivatives and so on – inverse effects in structure or in behavior. We propose quite formally to introduce for such comparisons the principle of direct unifying [41] and the alternative direct splitting (see Fig. 8.10d and context).

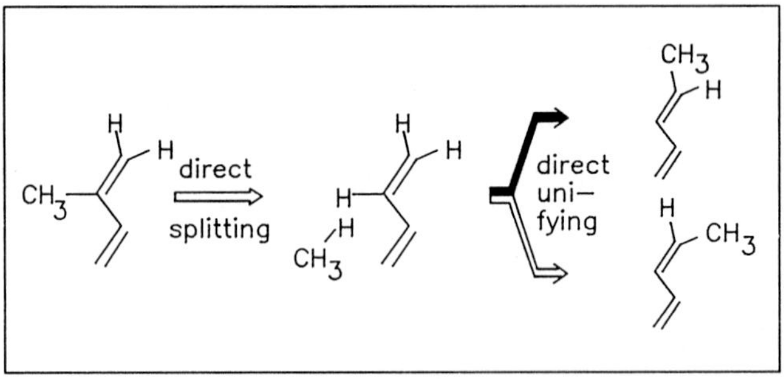

For a possible interpretation see in addition e.g. Fig. 3.21. It is demonstrated by four examples in Fig. 3.20 how identical perturbations at alternative positions in a chemical system bring about inverse effects on the states or processes. Some substituted 1,3-dienes are in thermodynamic equilibrium with their corresponding *cis*-1,2-divinylcyclobutanes [150]. (a): While the equilibrium of the head-to-tail dimerized four-membered ring with the two *trans*-piperylenes is located far on the side of the monomers, the dimer exceeds in the equilibrium of the isoprene [151]. Following (b) a rearrangement is observed to a norcaradiene-derivative, when both allylic positions in the cycloheptatriene are substituted by carboxylic acid ester groups. This does not occur, when the same substituents occupy the α,ω-vinylic positions [152]. The head-to-tail and tail-to-head as well as the tail-to-tail and head-to-head-coupling of methyl-substituted 1,3-dienes is by choice dependent in (c) both on the position of the methyl-groups (isoprene or *trans*-piperylene) and on the applied catalyst ($Ni^{(o)}$-P-ligand or 1,4-diazadiene-Fe) [153]. A more complex case is described in Fig. 3.20d. The asymmetric synthesis of 3-methyl-*E*-4-hexen-2-one is presented further in Fig. 1.2c [70,71] and will be discussed more in detail in Figs. 5.8 to 11. Phosphites with naphthyl- and (–)menthyl-groups are applied in a ratio 2:1 respectively 1:2 (alternative principles EVEN/ODD respectively ODD/EVEN) as ligands, which determine the optical induction in the β,γ-unsaturated ketone. The naphthyl substituent(s) is (are) coupled via oxygen alternatively in α-/β-position to the phosphorus. The combinations of α-position and 1:2 ratio of substituents or β-position and 2:1 ratio of substituents lead to a cooperativity to extreme values of optical yield in the ketone: $+92°/-16°$. The two other combinations effect compensation phenomena [154]: $-8°/+12°$ (for interpretation and definition see Fig. 4.5).

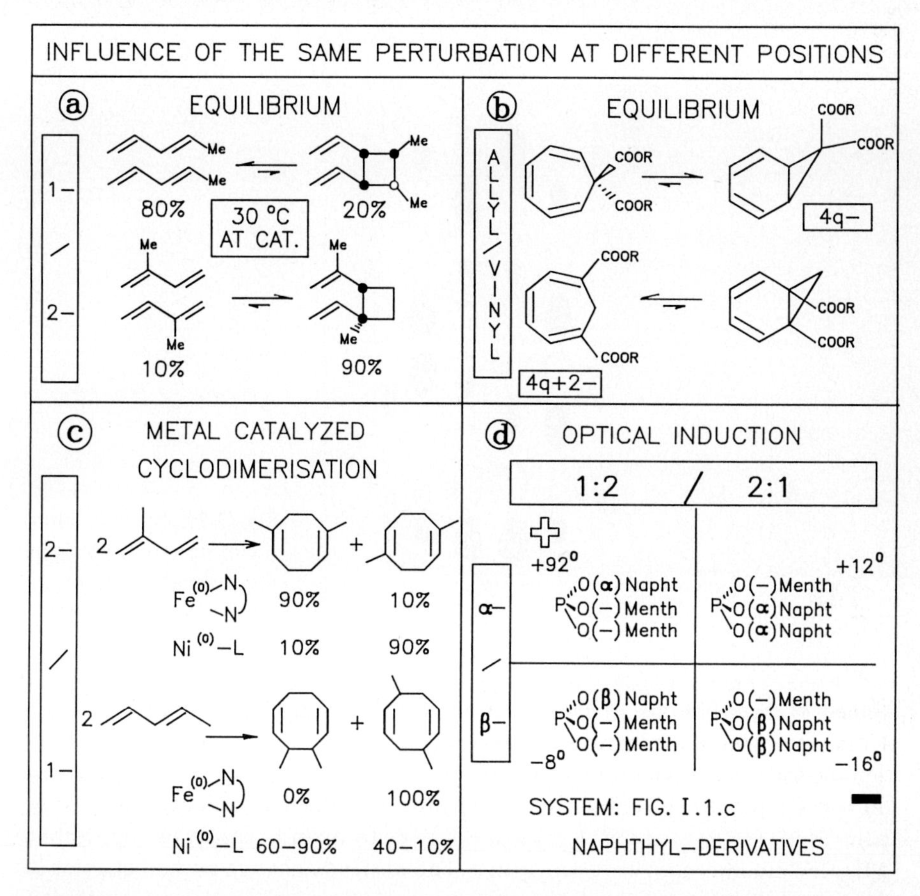

Fig. 3.20a-d. Four examples illustrating the importance of position for identical perturbations in equilibria in **(a)** and **(b)** and in Diels-Alder reaction **(c)** as well as in **(d)** in metal-induced asymmetric synthesis (for system **(d)** see Fig. 1.2c and Fig. 4.5d)

From the example in Fig. 3.21, it is clear, how the perturbations by the methyl group cause the destabilization of MO's of inverse symmetry in isoprene and piperylene [155]. A possible consequence is the fact, that e.g. S/AS coupling of two isoprenes or two piperylenes is alternatively observed at $Ni^{(o)}$-P-ligand – respectively 1,4-diazadiene-Fe-catalysts. See in addition Fig. 5.5.

3.8 The Alternative Principles IONIC/COVALENT

An extraordinary convenient example for the PSE-sector rule 1 is offered in Fig. 3.22a. A concerted reaction pathway *b* (with mainly covalent interactions) or a reaction pathway *a* via ionic intermediates is overwhelmingly populated by the

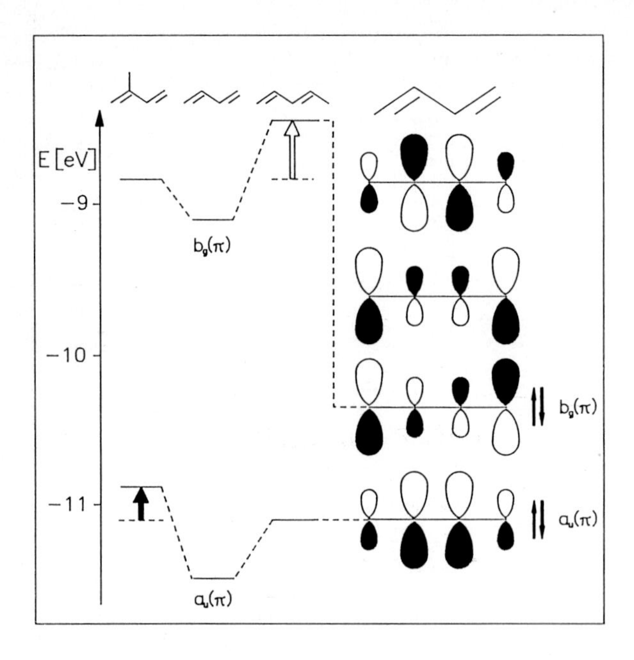

Fig. 3.21. Special additional destabilizations in the orbital energies of piperylene and isoprene

choice of heteroatoms in two positions of the catalyst in the catalyzed rearrangement of a disubstituted bis-homocubane derivative [156]. In order to demonstrate the cooperativity in both series of heteroatoms we have presented the relative populations of the reaction pathways *a* and *b* in form of a matrix. The pairs Sb + Cl (to *b*) and P + I (to *a*) far away from and near by the borderline AII/AIV are alternatively cooperative. The unsteady change in the behavior of the system by the choice of heteroatoms is also evident in the reaction rates likewise investigated [156]. By this a minimum is passed during change of the type of overwhelming interactions (negative vulcanic curve).

Peterson- [157] and Wittig-olefination are compared in Fig. 3.22b to demonstrate the alternative decision of stereochemical order (*cis-/trans-*olefine formation), firstly just by the sign of charge (H^-/H^+) and secondly by multi-dual decision trees of ALTERNATIVE PRINCIPLES (vide Chap. 5). The unequivocal control by charge in the Peterson-olefination is surely valid only, when all other ALTERNATIVE PRINCIPLES in the subsystems are slaved. *A statement of general validity for cis-/trans-control by charge ($C = +1/-1$) is forbidden and strictly dependent on system.*

Fig. 3.22a. Population of reaction pathway with overwhelmingly ionic (path *a*) respectively covalent interactions (path *b*) by the choice of heteroatoms in the catalysts depending on the position in the PSE. **(b)** Charge (Peterson-) and phase-relation control (Wittig-olefination) in two examples

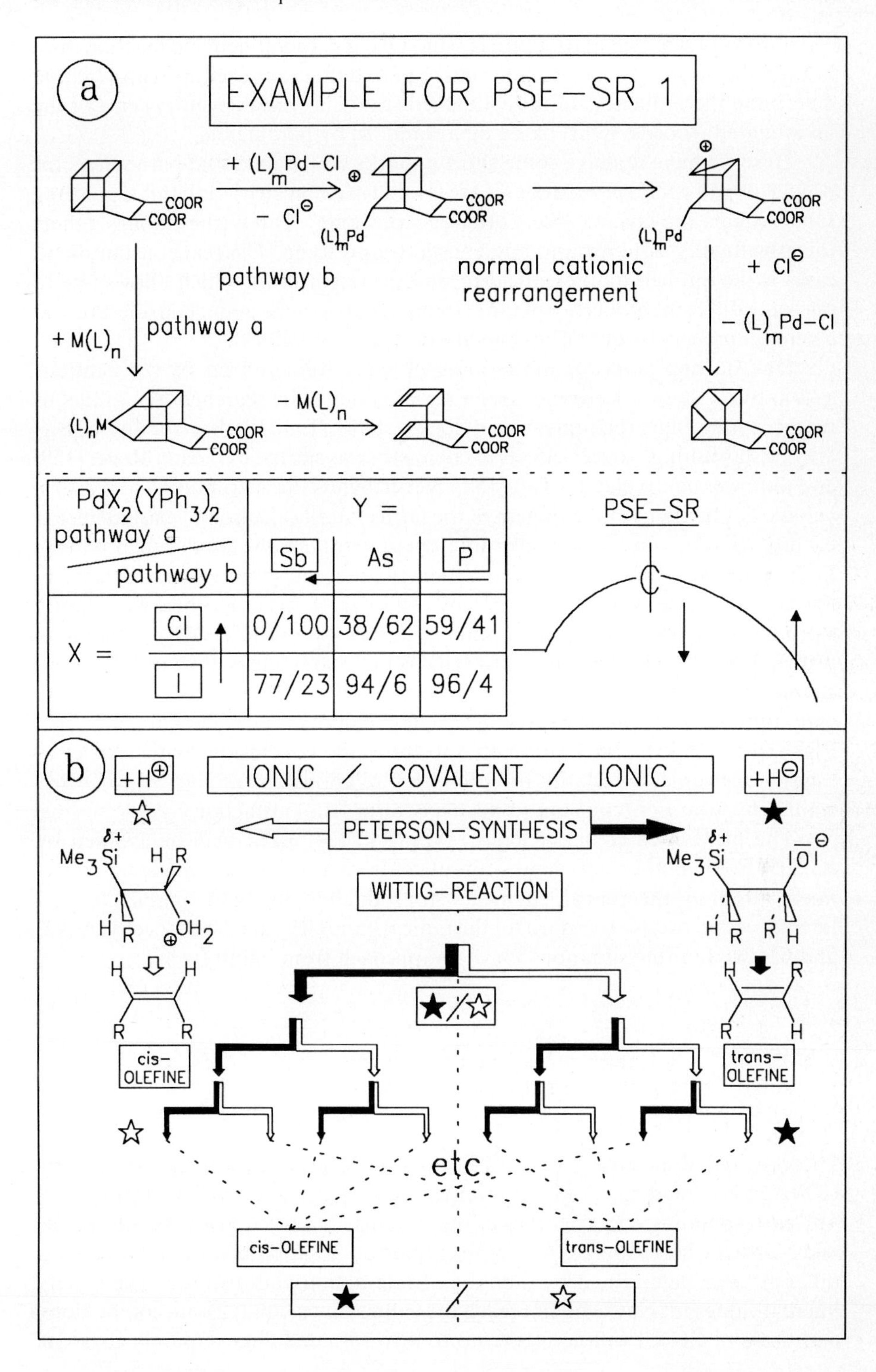
a
EXAMPLE FOR PSE–SR 1
+ (L)mPd–Cl
– Cl⊖
COOR
COOR
(L)mPd
pathway b
normal cationic rearrangement
+ Cl⊖
– (L)mPd–Cl
+ M(L)n
pathway a
(L)nM
– M(L)n
PdX2(YPh3)2
pathway a / pathway b
Y =
Sb
As
P
X =
Cl
I
0/100
38/62
59/41
77/23
94/6
96/4
PSE–SR
b
+H⊕
IONIC / COVALENT / IONIC
+H⊖
PETERSON–SYNTHESIS
WITTIG–REACTION
δ+
Me3Si
H R
H R
OH2
IOI
cis–OLEFINE
trans–OLEFINE
etc.
cis–OLEFINE
trans–OLEFINE

Two very famous catalysts are H^+ and OH^- e.g. for influencing reaction rates in saponifications. These two alternative possibilities are used by Taft [158], to determine the influences of individual substituents (variable subsystems of the investigated whole system) based on a standard by parameters.

Here we have to make some short remarks on the fact, that parameters for properties of systems or subsystems are termed in chemistry – and this is puzzling for beginners and laymen – very often as "constants". This is true amongst others for substituents, equilibrium, rate constants and so on. The real constant in all cases is the applied mathematical formalism (algorithm), which allows one to mark the different properties of the systems or subsystems under consideration in a general process by quantities (stochastic averaged values).

One starting point for the terming of space filling effects by the fictitious alternatives "steric/electronic" (see Fig. 3.23) has to be searched in a model of receptor-messenger relations as a key-lock system of Emil Fischer (for this see Fig. 8.12). Establishing "steric" effects in chemistry was started by Victor Meyer [159] and then was completed by Taft [158]. Nevertheless there are sufficient reasons, to have doubt about the setting up of the terms "steric/electronic" parameters as e.g. E_S^C and σ by considering only differences in transition states (pars pro toto). In the same way, one could assume arbitrarily that the whole cascade of steps in the saponification was perturbated both by interactions of a center with empty s-orbitals (H^+) and by those of a center with filled p-orbitals (OH^-) (alternative principle S/AS). Thereby the "steric/electronic" parameter sets would represent characteristic values for S/AS perturbations. These are burdened in addition with blindmaking, information destroying compensation phenomena (vide Figs. 2.12 and 13). One could come to the same conclusions with respect to "steric/electronic" effects by the presented relationships of Fig. 3.12b (AS/S control by term I of repulsion respectively term III of attraction).

The mentioned errors of logical typing in Fig. 3.23a may be avoided by starting from another statement without contradictions. In an imaginary experiment (and in theoretical chemistry reproducible by mathematical experiments) we choose as a standard for the radical pairs CH_3• and Cl• respectively Na• and Cl• two limiting situations for their approach from infinity:

(1)	$CH_3 - Cl$	and	$Na - Cl$
	0,5 0,5		0,5 0,5
(2)	CH_3^+ Cl^-	and	Na^+ Cl^-
	0,0 1,0		0,0 1,0

The considered electron pair (1,0) is equally distributed to both partners for COVALENT bonding or only contributed to one partner for IONIC interactions. The corresponding energy curves of the systems are obtained for the attraction, and – on a too close approach – for the repulsion, by approaching and following different laws determined by the type of interaction and distance. The experimental values for NaCl will not reach the values for calculated ionic interactions and those of CH_3Cl will not come up to the estimated data of purely covalent

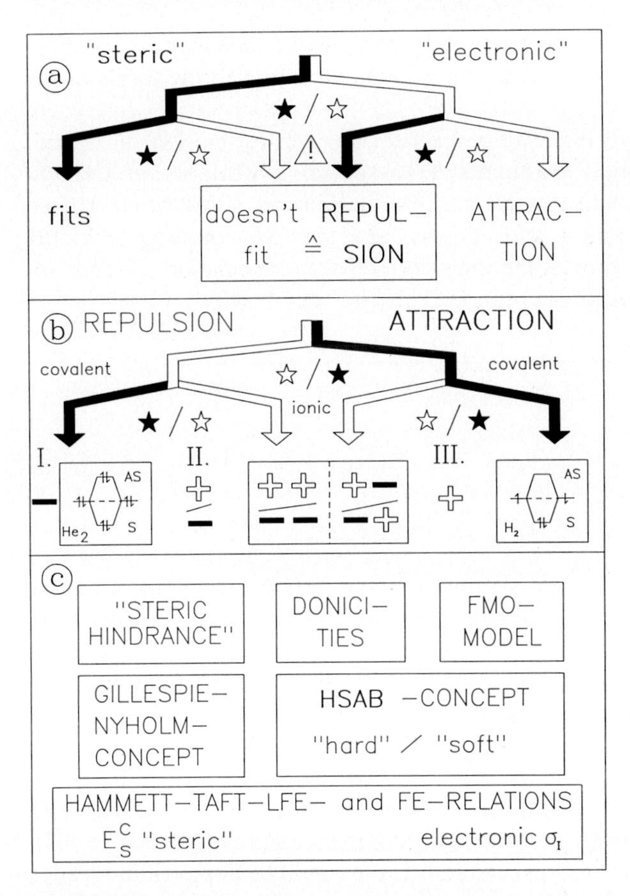

Fig. 3.23a. The contradiction in "steric/electronic" interactions with the alternatives IT FITS/IT DOESN'T FIT in key/lock relations and ATTRACTIVE/REPULSIVE in electromagnetic interactions. **(b)** Double-dual differentiations in electromagnetic interactions without errors in logical typing. **(c)** Relations of individual models to the three terms of the Klopman-Salem theorem

interactions [160]. The simple double-dual bifurcations for the factors REPULSIVE/ATTRACTIVE and COVALENT/IONIC interactions correspond to the three terms of the Klopman-Salem theorem [43] (Fig. 3.23c). In (b) the importance for the three terms of the theorem is represented in an oversimplified manner for beginners: (I) He_2 doesn't exist! in the ground-state; (II) Identical charges repel and opposite charges attract; (III) H_2 exists as a molecule with two atoms!

The highest occupied orbitals of the stable H_2 and the unstable He_2 are inverse with respect to their symmetry (S/AS). This last statement, in particular, should arouse interest with respect to the declarations and the "scope of validity" for different reactivity models (vide Fig. 3.23c) in relation to the terms of the Klopman-Salem theorem.

However it is noteworthy with respect to the over-estimation of "steric" effects, that space filling substituents have special advantages in CHIPS-chemistry. These bulky substituents support stability in CHIPS-perturbated systems (vide Figs. 3.1 and 4 and context) and the C-C-coupling of highly alkylsubstituted carbon atoms sometimes even results in special benefits e.g. for selenium, tellurium or silicon reagents [161, 162].

O + SeMe Li —ether, 77%→ HO SeMe

Even unusual stereochemical arrangements in CHONX-chemistry, e.g. tetrahedrane, are stabilized by tertiary-butyl groups (called "steric" shielding). In every case one has to consider besides the repulsion, that the σ-/π-separation breaks down in highly branched σ-systems. Theoretical chemistry is obliged to correct some of the usual assumptions about "steric" interactions [81].

In this chapter we have presented a whole series of ALTERNATIVE PRINCIPLES. But the series is by no means exhausted. So far, we have not considered e.g. END ON/SIDE ON (=180°/90°) [163] or CONCAVE/CONVEX [164] and other ALTERNATIVES of the second type (vide Fig. 8.2), which are so important for an experimentalist. We would be happy, if the reader would inform us of further general examples.

4 Representation of Differentiation and Compensation of ALTERNATIVE PRINCIPLES

Complicated: Nothing is more difficult than simplification. Nothing is simpler than complication.

G. Elgozy
Esprit des mots – l'antidictionnaire

We became aware of the ordering CONCEPT OF ALTERNATIVE PRINCIPLES more and more, when we – instead of presenting quantities in tables – went over to relating pairs of quantities to each other [29] (vide Figs. 4.5 and 6). But first of all we would like to consider two ALTERNATIVE PRINCIPLES and the possibilities of presenting their inverse effects.

4.1 On the Definition of Paritetic and Complementary ALTERNATIVE PRINCIPLES and their Effects

What we mean by our definitions of paritetic and complementary ALTERNATIVE PRINCIPLES is most easily clarified in the case of optical induction (real examples see Figs. 5.7 to 10). It is necessary, in the synthesis of enantiomeric compounds, to introduce somewhere in the educts and the effectors – or subsystems thereof – one of the two possible parities $P = +1/-1$. When this parity in a molecule is broken just once as shown schematically in Fig. 4.1, the sign of the specific rotation changes in the product by choosing the inverse parity. But the extent of the enantiomeric excess formed is – within the experimental error – absolutely constant and the sign is fixed. Additional variations such as E/Z-configuration in one of the educts (vide Fig. 5.6) might result in a change in the corresponding enantiomeric excess in the product mixture, too. Such ALTERNATIVE PRINCIPLES we call complementary, because these alternatives in the molecules lead to states with similar enthalpies but not yet to energy degenerated states (see Fig. 10.1). The com*p*lement*arity* includes the *parity* in latent manner in subsystems, but in addition we have to consider enthalpy differences and take them into account.

A consequence of this additional difference in enthalpy is, that the enantiomeric excesses in the products may be still related absolutely in some cases to the alternative parities, but the extent (the quantities of the optical induction) is different. When the effects of complementary ALTERNATIVE PRINCIPLES are very small in the system, very often only one type of enantiomeric excess is observed in the product even after changing these ALTERNATIVES; in this case only the quantities (and not the signs) are different: a relative effect of a relative

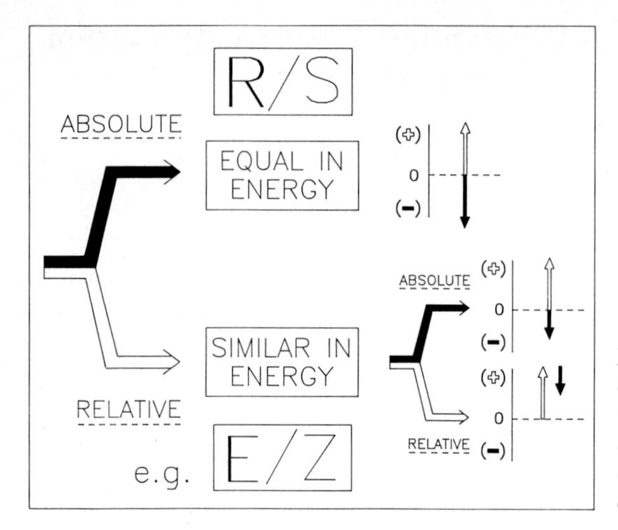

Fig. 4.1. Effect of paritetic (e.g. *R/S*) and complementary alternatives (e.g. *E/Z*) on direction and extent of optical induction for example

control by complementary factors. It is of special advantage in the representation of these results for these dominating, cited cases in the real examples, not to allow the important qualitative difference in the effect to vanish (see Scheme 4.1).

Instead of, as for example in (a) being shown by two values with the same signs, the quantities of the same preponderant behavior are averaged and represented by two opposing black and white arrows (b). Taking the variations in enthalpy into consideration by choosing a new standard from which only the latently present parities display their effects absolutely, this may be represented in the form of vectors with positive and negative directions (c) or even as opposite vectors starting from this new standard (d). Another possibility is to start from this new standard with bifurcations (e).

The examples with rules of *absolute* inverse order are extremely rare and the assumed regularities may be applied in a more general manner as *relative* rules. This we would like to prove by means of the so-called lacton rule for derivatives of sugar acids, which was formulated at first by Hudson as an *absolute* and later on by Freudenberg as a *relative* rule intended to broader application. This rules are expected to correlate directly the stereochemistry at the stereogenic center of the lacton function with the values of the specific rotation. Accordingly, absolutely formulated: When the lacton oxygen is "on the left" in the Fischer projection, the sign of the rotational value is negative and vice versa. Accordingly, relatively formulated: When the lacton oxygen is "on the left" in the Fischer projection, the rotational value of this enantiomer has always a smaller value on the scale (greater negative or smaller positive) and vice versa [165].

Nevertheless we should add, that inverse rules could be valid. This could result e.g. by ATOM REPLACEMENT O/S in lacton position or only after total change to the thioester (–COOR/–CSSR) or even only in the all-thio-sugar derivatives.

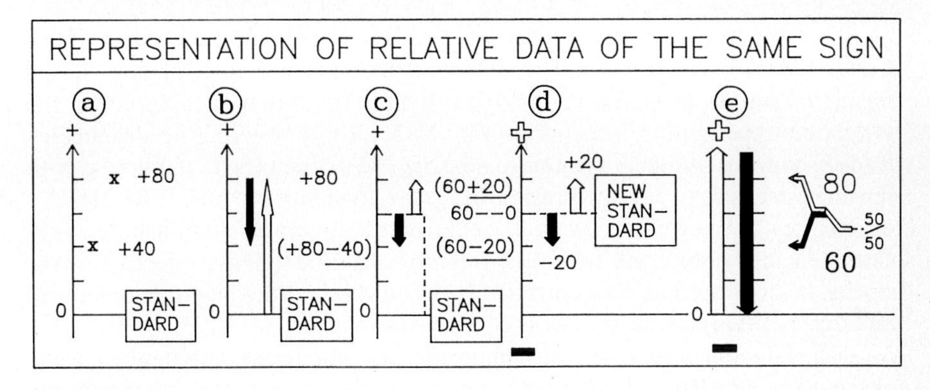

Scheme 4.1

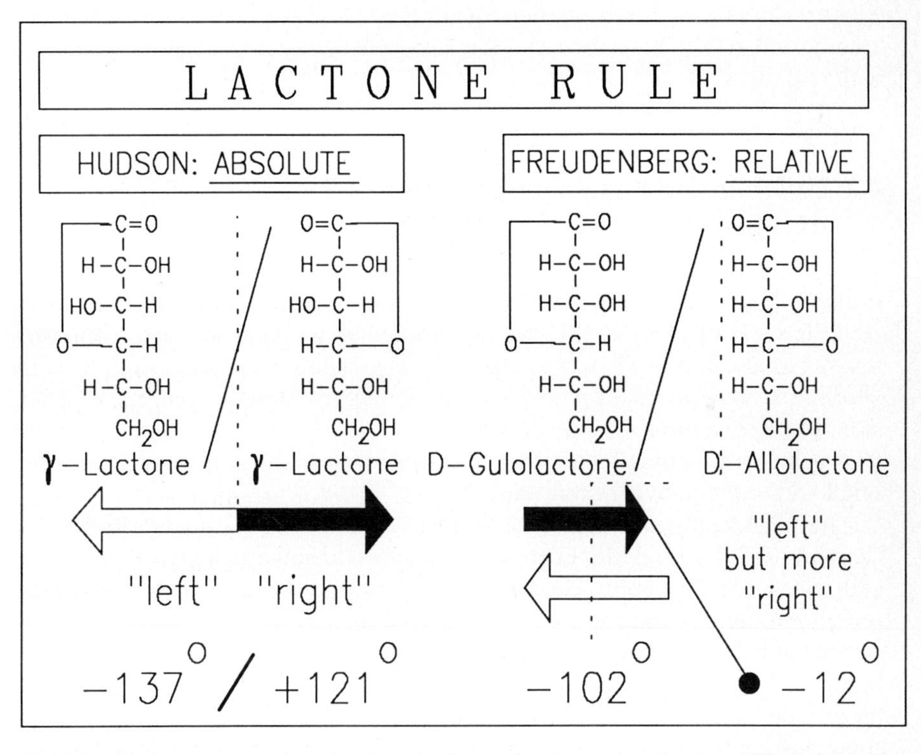

Fig. 4.2. Concrete examples obeying the lactone-rule in the absolute (Hudson) and relative formulation (Freudenberg)

Two important aspects necessary to the fundamental understanding of this CONCEPT OF ALTERNATIVE PRINCIPLES are presented in Fig. 4.3 by means of simple mechanical analogies. It may be clearly recognized by the sympathetic pendulum in (a) (vide Fig. 10.2, too), that giving up degeneracy leads in coupled swinging systems to *time dependent* changes in amplitudes with specific directions of energy transfer. In (b) the giving up of "LEFT/RIGHT" – degeneracy of the whole system at the Bernoulli apparatus demonstrates, very simply in an absolute and relative manner by absolute/relative effects, how the stochastic distributions are controlled by multi-dual (at the beginning identical and degenerated) decision trees as compared to Fig. 4.1.

Fig. 4.4a and b represent schematically, that the four combinations of the alternatives STABILE/LABILE for structures (and compounds, too) with the alternatives ACTIVATION/INHIBITION (here shown by the same values) for processes (reactions) automatically give rise to a MAXIMUM/MINIMUM (positive and negative cooperation) and two compensations of the product-stream. The observed combination e.g. for the MAXIMUM is system dependent as indicated in (c). These abstract discussions will be easier to understand later on with real cases.

4.2 Differentiation and Compensation of Two Pairs of ALTERNATIVE PRINCIPLES

In the following examples aimed to elucidating the interactions of two pairs of ALTERNATIVE PRINCIPLES, we proceeded in the completely opposite manner as in Chap. 3. There we presented only the less numerous examples with absolute inverse effects of pairwise ordering factors. Here in contrast, we have selected those examples, for the following figures, which are not mentioned in the textbooks, because the effects – individually and separately considered – are very small and only relatively expressed.

In Fig. 4.5a the interdependent influences of Na^+/K^+ as counterions of the "carbanions" [166] and Me/Et groups in the esters are investigated in the Michael addition under coherent conditions (under argon): rate of C-C-coupling (measured here by the conversion). One combination (Na^+ + Me) is superior compared to all three others [122]. See further below the detailed investigations in Figs. 5.3 to 5. The influences gained by choosing the alkali metals Na/K and the number of methylene groups in the di-esters is investigated in the Dieckmann condensation by the conversions in five- or six-membered ring syntheses in Fig. 4.5b. Perhaps it would be advantageous in all these cases to abolish the term "carbanion chemistry" and to respect the influence of the counterions by the term metalorganics. In (c) the control exerted by –OMe/–SMe in *ortho-/para*-position of the phenyl-group in the aldehyde was investigated in the allphenyl Wittig system (with three phenyls at the phosphorus and one phenylsubstituent both in ylide and aldehyde position) taken as standard (42% *cis*-selectivity) in the four

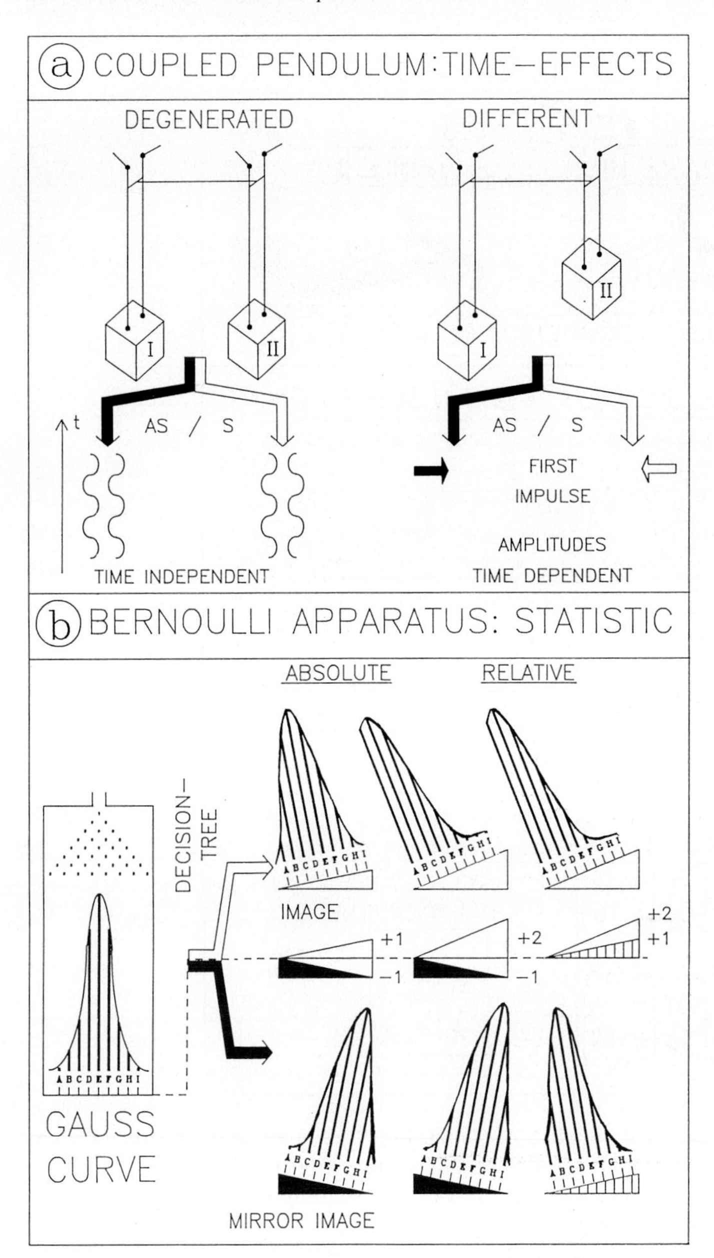

Fig. 4.3a,b. Simple mechanical models: **(a)** The sympathetic pendulum points to time dependent preference of alternative dynamic processes with respect to symmetry (S/AS) by giving up degeneracy. **(b)** The Bernoulli apparatus is a simple mechanical model for the considerations in Fig. 4.1

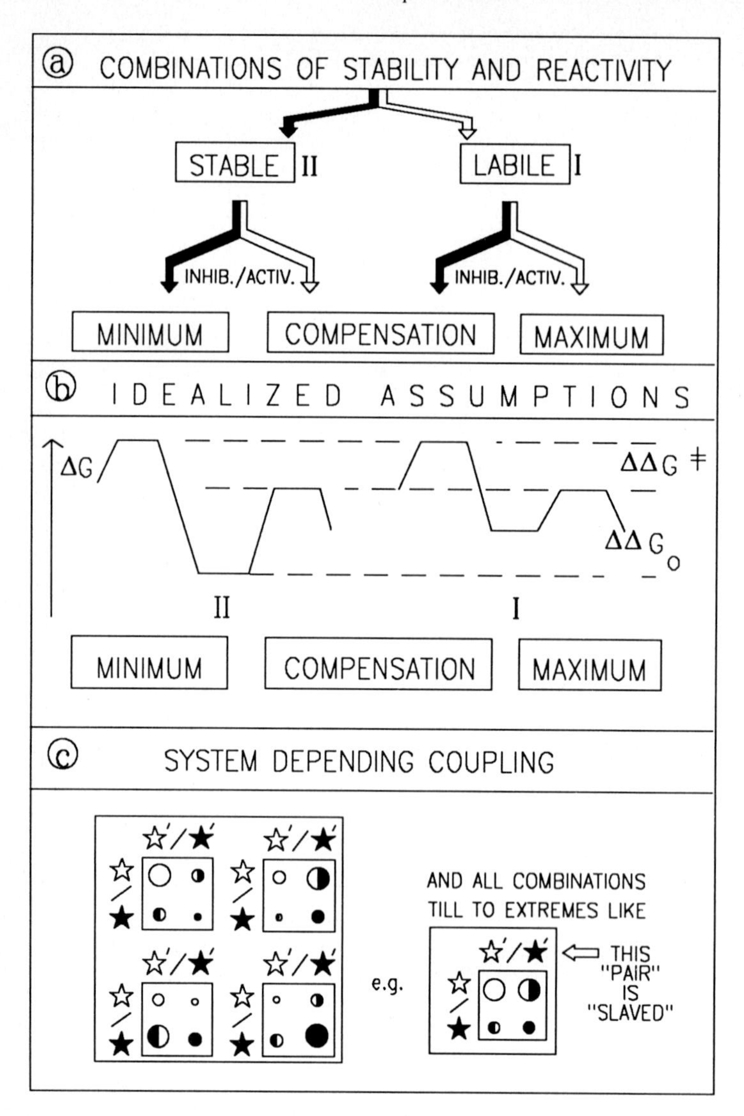

Fig. 4.4a-c. The simplified coupling of stabilization/destabilization (in structures and states) with activation/inhibition (in processes) is represented in **(a)** as usual in chemistry and schematically in **(b)**. **(c)** shows e.g. the effects depending on system on the product streams to alternative structures

combinations on the *cis*-selectivity of the correspondingly substituted stilbenes. An –OMe group in alternatively *ortho-/para*-position causes an absolute effect with respect to the standard; in the case of an –SMe substituent only a relative control is observed [30].

The most abstract example is demonstrated in Fig. 4.5d with respect to the CONCEPT OF ALTERNATIVE PRINCIPLES. It was investigated in the asymmetric Ni-induced synthesis of 3-methyl-*E*-4-hexen-2-one (vide Figs. 1.2c and 5.7 to 10), how the type of direct unifying in α- or β-position of naphthyl groups via oxygen with the central atom of the P-ligand and the ratio (2:1 and 1:2) of these naphthyl subsystems to the parity breaking –O-(–)-menthyl groups in the corresponding phosphites control the sign and the value of the specific rotation in the ketone. From this the individual enantiomeric excess can be calculated. The choice of α-/β-position determines the sign (absolute control) and the ratios of substituents the extent of optical induction (relative control). Both pairs are complementary ALTERNATIVE PRINCIPLES. The parity of the INPUT information is determined in all cases by the number of (–)-menthyl groups [154].

The complexity in Fig. 4.6 increases steadily in the pairwise comparison of the effects by constant classes of alternatives. In the example (a) two classes of ALTERNATIVE PRINCIPLES in *n*-propyl-/phenyl-groups alternatively in $R^3 + R^2/R^2 + R^3$ positions of a Wittig-system were investigated in their effects on the *cis*-selectivity of 1-phenyl-pent-1-ene by differentiation and compensation with the classes of ALTERNATIVE PRINCIPLES in the salts LiI/$MgBr_2$. The salts influence, in every case, alternatively the *cis*-selectivity in the systems with positional alternatives. The salt-free positional alternatives yield the following *cis*-selectivities: $R^3 + R^2 = n\text{-Pr} + \text{Ph} = 33\%$ respectively $R^2 + R^3 = n\text{-Pr} + \text{Ph} = 94\%$ [31]. In case (b) the interactions of the alternative principles MUCH/A LITTLE (see Chap. 6, Fig. 6.12 and 13, too) with those constant classes of ALTERNATIVE PRINCIPLES in N- (morpholine) and P-ligands (triphenylphosphite) is described by the selective synthesis of dimers of butadiene with alternatively CYCLIC (*cis, cis*-1,5-cyclooctadiene) and OPEN structure (*n*-octa-1,3,6-trienes) at $Ni^{(0)}$-catalysts [13].

4.3 Alternative Patterns by Classes of ALTERNATIVE PRINCIPLES

Pattern comparison is superior to columns of numbers in a table as distinctly demonstrated in Fig. 4.7. The chosen example was very carefully investigated by Ashby [167] and some results are published in the presented tables. Important evidence (relations) can be obtained, of course by separating the two tables as demonstrated. The effect of an alkyl variation (Me/Et/*i*-Pr/*t*-Bu) in a Grignard reagent on the relative reaction rates with acetone or benzophenone (separated in the table) is investigated. The corresponding inverse trend in the values can be recognized immediately in both series. But the kind of representation – chosen

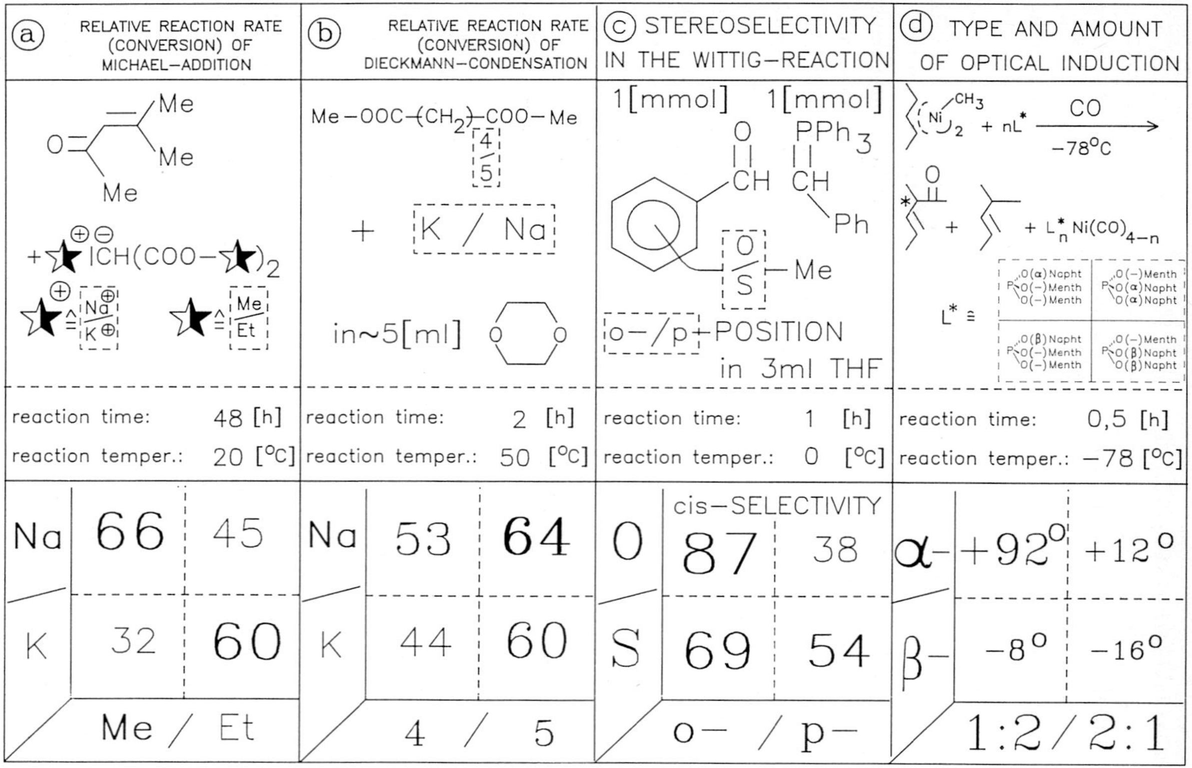

(a) RELATIVE REACTION RATE (CONVERSION) OF MICHAEL–ADDITION	(b) RELATIVE REACTION RATE (CONVERSION) OF DIECKMANN–CONDENSATION	(c) STEREOSELECTIVITY IN THE WITTIG–REACTION	(d) TYPE AND AMOUNT OF OPTICAL INDUCTION
Me, Me, Me; + ★ ⊕⊖ ICH(COO–★)$_2$; ★ ≙ Na⊕ / K⊕; ★ ≙ Me / Et	Me–OOC–(CH$_2$)$_{4/5}$–COO–Me + K / Na in ~5[ml]	1[mmol] CH=O, 1[mmol] PPh$_3$=CH–Ph; O / S –Me; o–/p– POSITION in 3ml THF	$(Ni–CH_3)_2$ + nL* →(CO, –78°C) + $L^*_n Ni(CO)_{4-n}$; L* ≙ P(O(α)Napht)(O(–)Menth)$_2$ / P(O(–)Menth)(O(α)Napht)$_2$ / P(O(β)Napht)(O(–)Menth)$_2$ / P(O(–)Menth)(O(β)Napht)$_2$
reaction time: 48 [h]	reaction time: 2 [h]	reaction time: 1 [h]	reaction time: 0,5 [h]
reaction temper.: 20 [°C]	reaction temper.: 50 [°C]	reaction temper.: 0 [°C]	reaction temper.: –78 [°C]

(a)	Me	Et
Na	66	45
K	32	60

(b)	4	5
Na	53	64
K	44	60

(c) cis–SELECTIVITY	o–	p–
O	87	38
S	69	54

(d)	1:2	2:1
α–	+92°	+12°
β–	–8°	–16°

Fig. 4.5a-d. Four examples for the cooperativity of two pairs of ALTERNATIVE PRINCIPLES concerning reaction rates in two different reactions in **(a)** and **(b)**, *cis*-selectivity in **(c)** and type and extent of optical induction in **(d)**

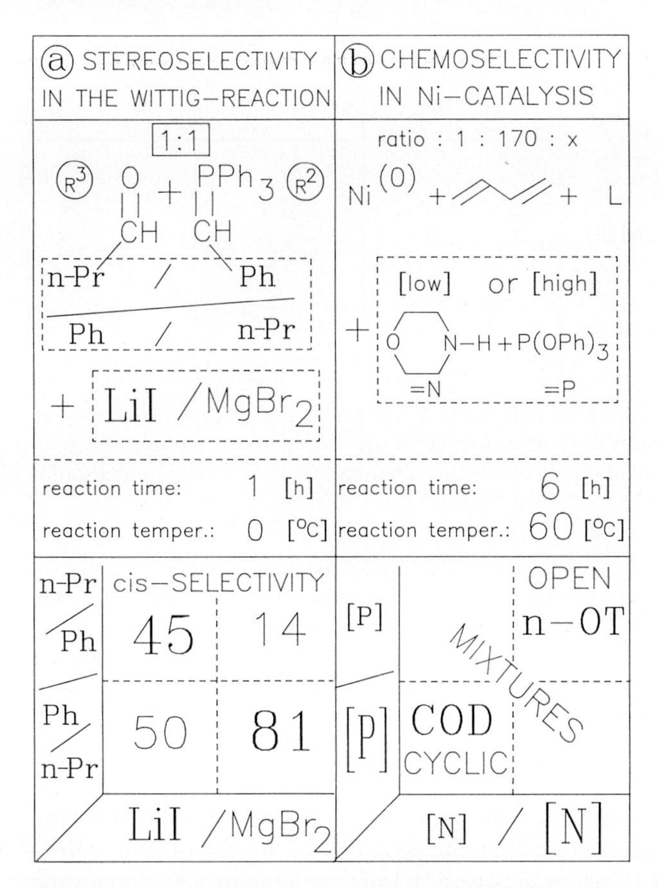

Fig. 4.6a. Not identified pairwise classes of ALTERNATIVE PRINCIPLES in salts ($LiI/MgBr_2$) and such in substituents (*n*-Bu/Ph) depending on R^3/R^2-position in the Wittig system show phenomena analogous to Fig. 4.5c for *cis*-selectivity. **(b)** The ALTERNATIVE PRINCIPLES of high/low concentrations (MUCH/A LITTLE) are coupled with those in DO/ACC at the central atom of the ligand at the metal

by us – in the form of patterns in the left hand side of Fig. 4.7 demonstrates quite distinctly the existence of two inverse alternating patterns. When a pattern is inverted one might anticipate, that at least one quality in the system or the process has to change. This other quality could be the suggested change in the reaction mechanism (SET/IONIC SPLITTING) proposed by the author [167].

What different types of patterns may in principal be identified by systematic variations at tetrahedral centers leading to four data, was presented in connection with Figs. 2.3 and 4.

Two well known alkyl series are presented once again in Fig. 4.8a and b in groups of four substituents: the homologous series as type I and the branching series as type II. In addition we propose the four variations in Fig. 4.8c, to reduce

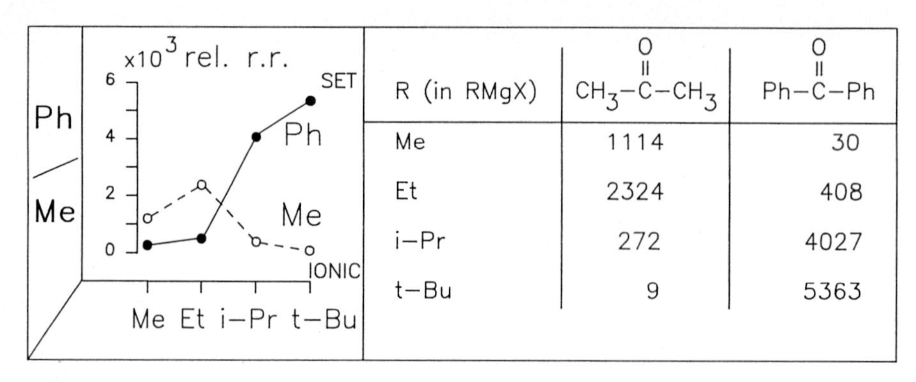

R (in RMgX)	$CH_3-C(=O)-CH_3$	$Ph-C(=O)-Ph$
Me	1114	30
Et	2324	408
i–Pr	272	4027
t–Bu	9	5363

Fig. 4.7. Representation in form of patterns is superior to listing in tables (see text). The dependency of reaction rates of Grignard reagents with alternatively acetone or benzophenone is controlled in inverse patterns by the variation Me/Et/*i*-Pr/*t*-Bu under standard conditions

the necessary experimental investigations for elucidating the effect of alkyl groups and on the other hand to be as certain as possible to comprehend all extreme influences of the alkyl groups (besides those of longchained alkyls in vesicles, micelles and membranes). Our proposal accounts for multifold reasons. On the one hand a twofold change in ALTERNATIVE PRINCIPLES has not to be feared by omission of the *tertiary* butyl group (see Chap. 2, the text after the discussion of Fig. 2.16). But it is far more important, that in every position to be compared, a standard substituent – the hydrogen – is normally available for the comparison of alkyl control in different positions or by alternative stereochemistry as we demonstrated by the case of 1,3-dienes in Figure 4.8d. Finally the selection of the alkyls Me/*i*-Bu/*i*-Pr is determined by the facts, that Nature has selected these alkyl groups (the secondary butyl group with a stereogenic center is not taken into consideration) during evolution e.g. in the α-amino carboxylic acids. In the series Me/Et/*i*-Pr the ethyl group having been excluded by evolution is replaced by the *i*-butyl group (vide Fig. 9.3). The P-parameters are very similar: $P_{Et} = 2{,}32$ and $P_{i\text{-}Bu} = 2{,}52$. In addition, we have always found up to now – and that is only demonstrated by one example in Fig. 4.9 [31] –, that the isobutyl group always embodies “the more perfect ethyl group” in controlling every type of alternative order.

4.4 Symmetric/Antisymmetric Coupling of Two Pairs of ALTERNATIVE PRINCIPLES

Starting with Fig. 4.10 we would like to discuss once again and to present, by different figures, in detail the different consequences of two types of coupling (symmetric/antisymmetric or cooperative/compensative or dominant/recessive)

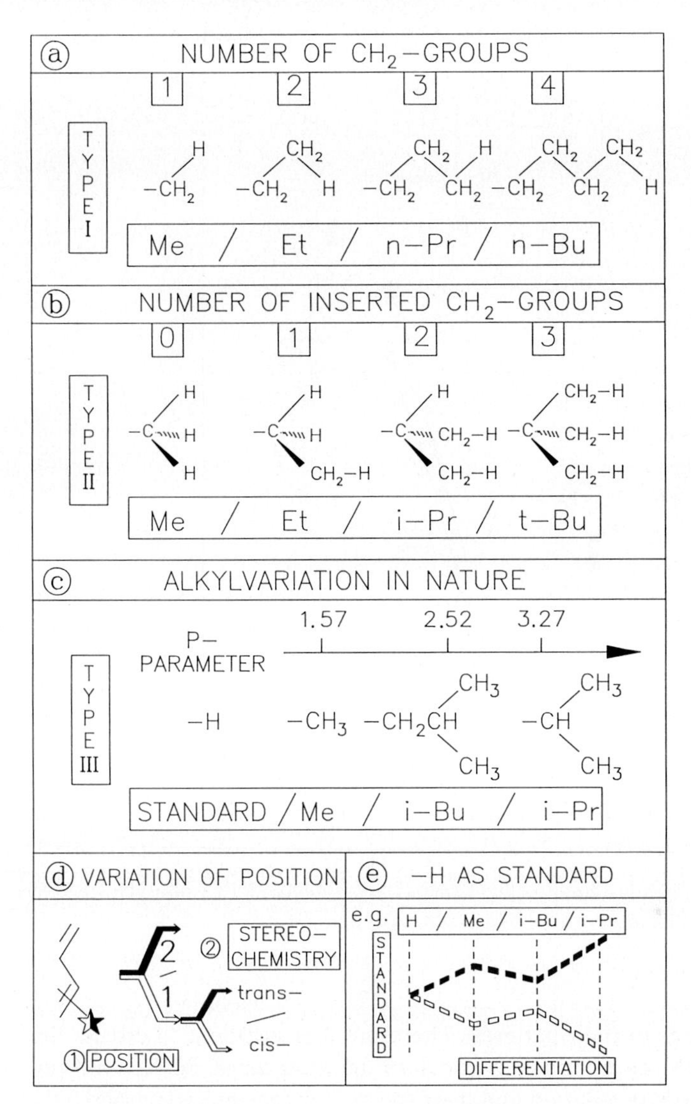

Fig. 4.8a-e. Systematic variations by EVEN/ODD numbers of methylene groups for homologous (**a**) and branching alkyl series (**b**). (**c**) The biomimetic variation H/Me/*i*-Bu/*i*-Pr is important for investigating e.g. the effect of 1-/2-position in 1,3-dienes in (**d**), because the hydrogen standard is valid for both positions (**e**)

by the example of ALTERNATIVE PRINCIPLES in the pairs *i*-Bu/*i*-Pr and Cl/Br influencing the shifts of the δ^{119}Sn-values in R_2SnX_2 and $R_4Sn_2X_2S$ compounds [168].

The δ^{119}Sn-values determined experimentally in R_2SnX_2 (a) and $R_4Sn_2X_2S$ (b) components are totally different with respect to their quantities, but offer

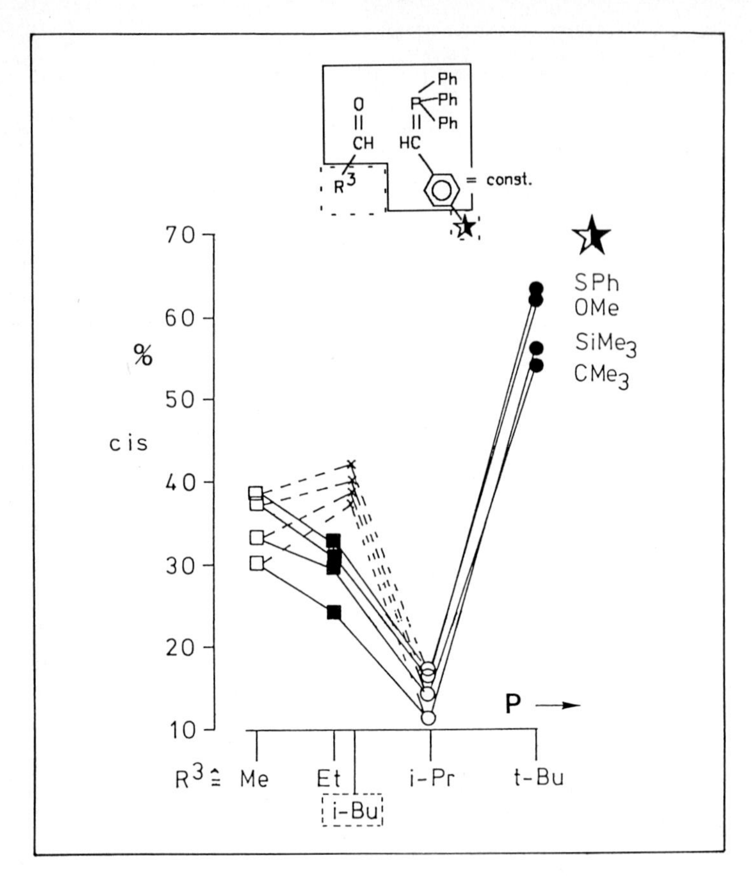

Fig. 4.9. Proof for the similar, but fortified control of *cis*-selectivity by isobutyl instead of ethyl groups in the aldehyde position of a Wittig system

surprising similarities in their patterns. Therefore it is sufficient to discuss the results of the R_2SnX_2 compounds, particularly because the $R_4Sn_2X_4S$ components dissociate partly in solution and their spectra become more complex.

We have offered, for understanding the interesting interactions based upon the data taken from a δ^{119}Sn-scale, two different patterns above and below this scale. In the pattern below the data both of the isobutyl (white) and the isopropyl subsystem (black) are combined, whereby in the first named series a strong differentiation is observed caused by the heteroatoms in the sequence Cl, Br, I as well known from the PSE and in the second named series a fargoing compensation leading to the unusual sequence Br, Cl, I can be identified. The pattern above the scale demonstrates quite clearly, that the influence of the *i*-Bu/*i*-Pr groups on the shift of the δ^{119}Sn-values is inverted by replacing Cl with Br. In the iodides the same pattern is observed as in the bromides. This alternation in the coupling of pairs of ALTERNATIVE PRINCIPLES is once again described

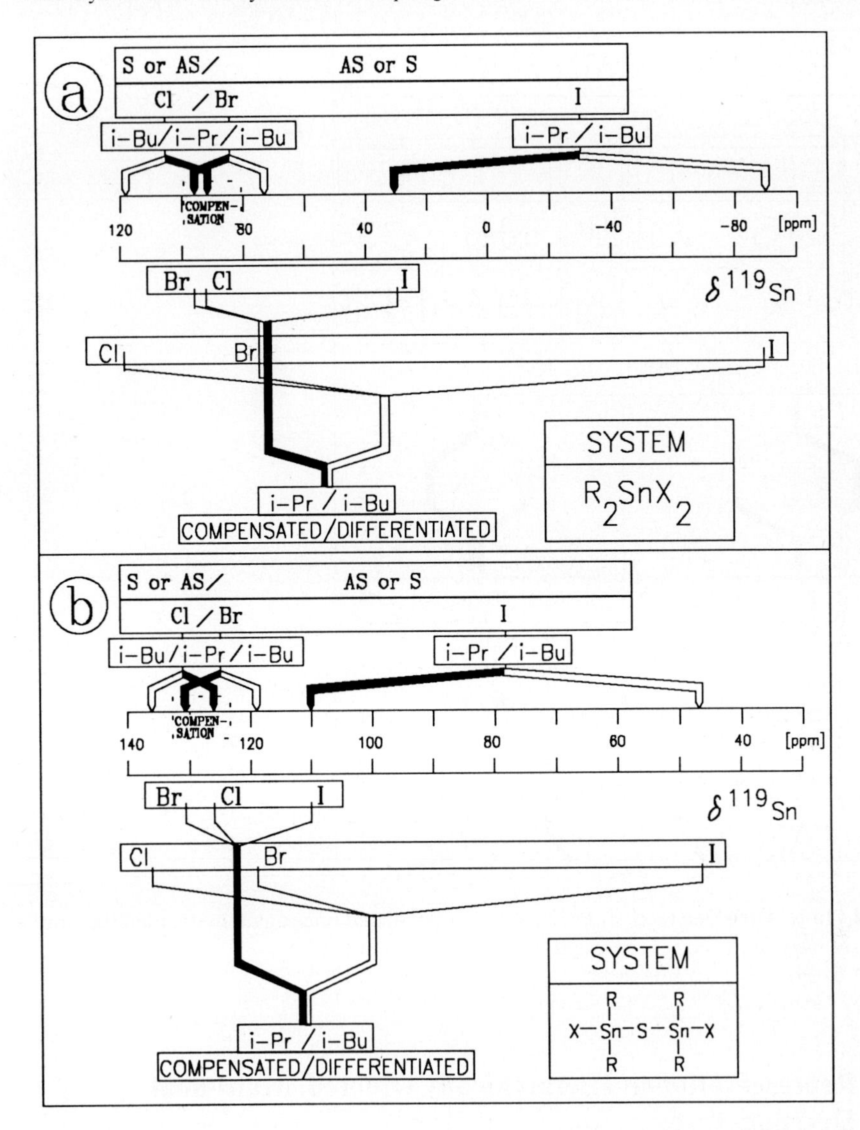

Fig. 4.10a,b. Important examples for the positive and negative cooperation or alternatively the two compensations of both ALTERNATIVE PRINCIPLES in the pairs Cl/Br respectively Br/I as well as *i*-Bu/*i*-Pr in two different systems [**(a)** and **(b)**] by spectroscopic data

differently in the Figs. 4.11a and b in the form of related pairs of quantities or as a double-dual decision tree above a scale, both representations making the same statement. In (c) the extreme values of couplings of a lot of ALTERNATIVE PRINCIPLES are depicted symbolically for evolved systems (for this see Fig. 9.1).

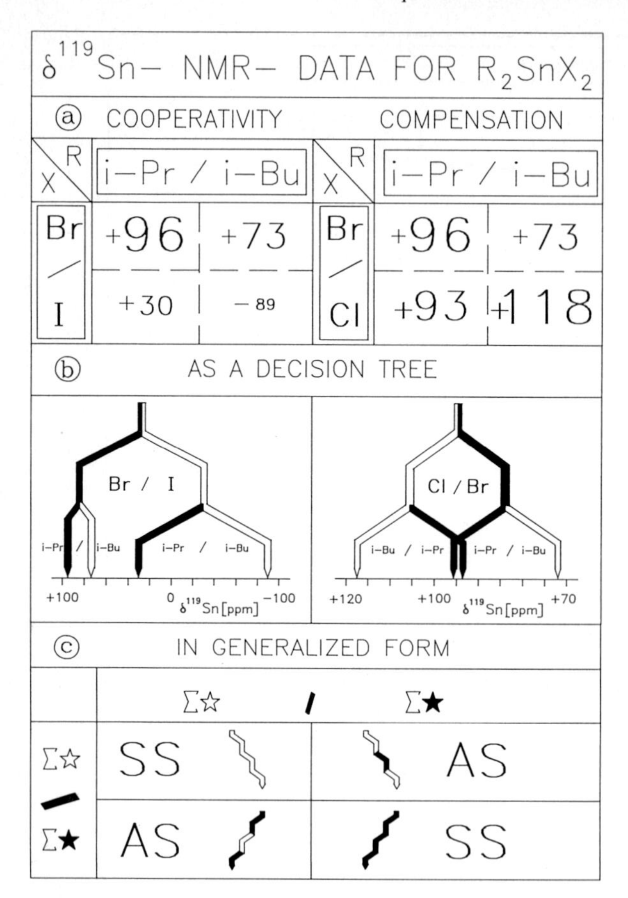

Fig. 4.11a-c. The same facts from Fig. 4.10a are depicted once again in detailed [(**a**) and (**b**)] and in generalized (**c**) representations of a different kind

4.5 Representation in Hierarchically Ordered, Multi-dual Decision-Trees

Investigating more than two pairs of ALTERNATIVE PRINCIPLES with relative small effects in chemical systems, which have to be evolved experimentally, representation by means of hierarchically ordered, multi-dual decision trees (bifurcations) proves to be extremely useful. The procedure of constructing such trees consisting of bifurcations shall be explained hypothetically by the facts placed together in Fig. 4.12.

Every addition of new ALTERNATIVE PRINCIPLE doubles the number of necessary experiments when they are introduced in any subsystem interacting with the parent system. Sixteen experiments have to be realized in this abstractly

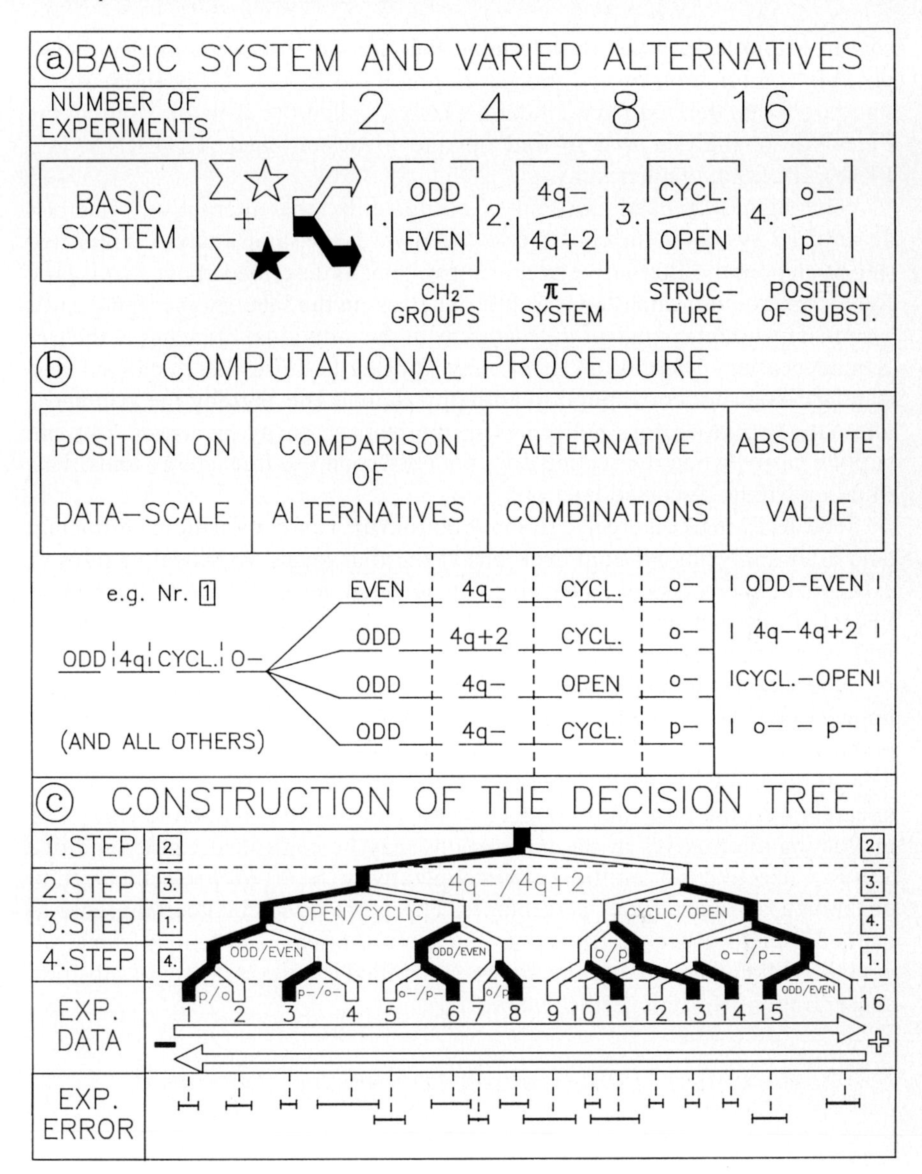

Fig. 4.12a-c. Four pairs of ALTERNATIVE PRINCIPLES **(a)** are systematically investigated in their influence on an alternative behaviour of a system in **(b)**. This analysis determines, step by step **(c)**, the corresponding ALTERNATIVE PRINCIPLES of the highest hierarchical order. The tree is constructed from the bottom to the top (see text)

chosen example in order to determine the effect of every variation on each possible combination by the procedure presented in Fig. 4.12b, to identify a standard value derived from all data, which is called the skeleton value in the Free-Wilson analysis, and to find out the highest ranked ALTERNATIVE PRINCIPLES in the hierarchy.

The highest ranked position is occupied by the alternative principles 4q–/4q+2-system in the chosen case. Afterwards the same analysis is repeated step by step for the alternative trees as subsystems as described above for all data. By this method our intention is to rule out errors in the signs by averaging and a system dependent change of the hierarchy in the individual branches of the tree is made easier. Thereby an error may appear, because the calculated new standard does not correspond to enthalpy change and thereby the symmetric alternative splitting does not represent the realistic parity inversion. This can happen easily, when the scaling does not correspond to free energy units. This situation will be discussed later on.

The decision tree is erected from the bottom up, based upon the experimental data after analyzing all dual decisions at the four levels. A critical analysis of experimental errors will determine, up to what level the ALTERNATIVE PRINCIPLES can significantly be discerned, when they overcome the interval defined by experimental error.

Special aspects have to be considered, when a multidual decision tree is built upon a percentage scale of yields or conversions being of special interest to the preparatively working chemist. The scales for these values should be represented logarithmically with respect to energy, because free enthalpy is always related linearely with the logarithm of the concentrations of any compound [169]. The interacting alternatives in the bifurcations may be controlled e.g. by additive (30–70% selectivity) or multiplicative cooperativity (≈ 90–98% selectivity) [170] depending on the range of percentages. This corresponds in a certain sense to defined pseudo orders (2,1,0) e.g. in the general Michaelis-Menten formalism with the Hill-coefficient 2, which may be identified in certain ranges of substrate concentrations (for this see Scheme 6.2).

5 Representative Examples of Multi-dual Decision-Trees: A Generalization of Phase Relation Rules

The pattern, that connects.

G. Bateson [9]

In this chapter some controls are introduced consciously – they follow the CONCEPT OF ALTERNATIVE PRINCIPLES and are exemplified by means of digital decision trees. However we will also select the other possibilities of presenting connectivities described in the preceding chapter to accentuate special aspects.

5.1 The Control of Elimination Reactions by ALTERNATIVE PRINCIPLES

The system presented in Fig. 5.1a corresponds exactly to the partial aspect worked out in Fig. 1.3a. The direct correlation to alternative acid catalyzed six-membered ring syntheses can be strengthened by the synthetic steps offered in Scheme 5.1. Starting from mevalonic acid the alternative regioselective elimination of water to isopentenyl- or dimethylallyl-oxidiphosphate offers a quite direct analogy to the systems with alternative elimination of hydrogen atom from a primary respectively secondary carbon atom as investigated by us. The equilibrium in the mixture of "active isoprenes" is balanced by enzymes. Both structural isomers can easily be dimerized head-to-tail-wise by enzymes at the two centers with nucleophilic or electrophilic character forming geranyl-oxidiphosphate. Stereospecific isomerization (inversion of stereochemistry) to neryl-oxidiphosphate allows e.g. the alternative formation of limonene or terpinolene [171].

We would like to emphasize, that certain enzymes are able to couple – despite the pronounced electrophilicity in the allylic position – these positions of identical reactivity in the higher homologues of "active isoprene". By this type of coupling of equally reactive groups (with apparently the same signs of the charges) e.g. tri- and tetra-terpene are formed in a type of convergent strategy. Perhaps these "tricks" [172] of Nature are based upon the predominance of phase relation controlled (covalent) interactions (see also Fig. 3.16c).

We have chosen the *tertiary* amylhalides, because in this case the hydrogen atom is eliminated alternatively from a primary or secondary carbon atom controlled easily by small effects. In addition, no rearrangement of the skeleton has to be apprehended even when a carbenium ion is formed as an intermediate.

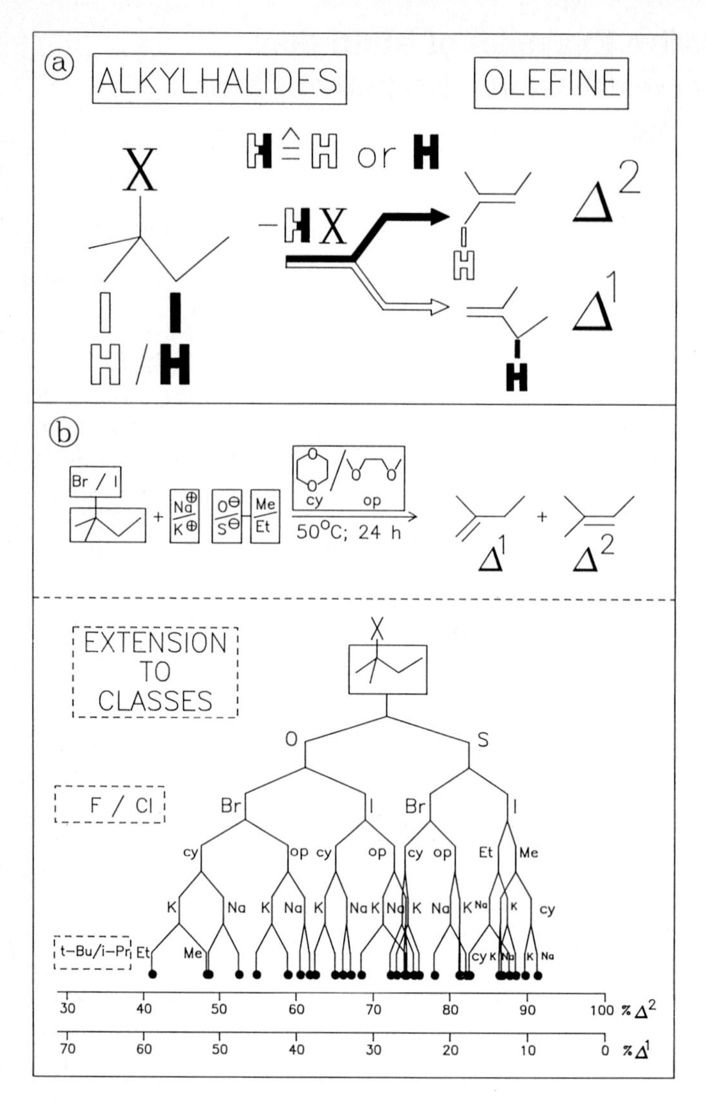

Fig. 5.1a. Δ1- or Δ2-olefines are formed from *tert*-amylhalides by HX-elimination. **(b)** The multi-dual decision trees for the bromide and iodide is constructed above the scale of olefine selectivities

The system allows only the alternative regioselective decision to a Δ1-/Δ2-olefine. A stereodifferentiation is not possible in the system (e.g. *cis-/trans-*).

All alternatives chosen by us have already been investigated, e.g. [173–176]. The unifying pattern of cooperativity and compensation in digital decision trees is shown in Fig. 5.1 for all pairs of ALTERNATIVE PRINCIPLES, which are not

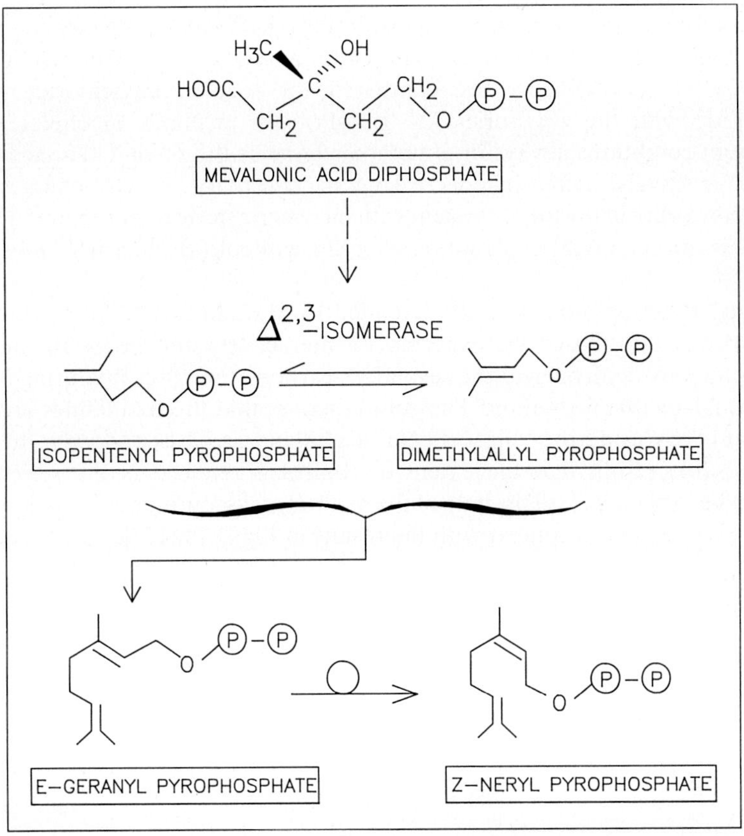

MEVALONIC ACID DIPHOSPHATE

$\Delta^{2,3}$-ISOMERASE

ISOPENTENYL PYROPHOSPHATE

DIMETHYLALLYL PYROPHOSPHATE

E-GERANYL PYROPHOSPHATE

Z-NERYL PYROPHOSPHATE

Scheme 5.1

yet present in the products (but in stoichiometrically inducing or catalyzing effectors and in the leaving group of the educt X = Br/I). Strikingly the strongest differentiations are effected by the choice of thioalcoholates or alcoholates (S/O) as bases. A system depending change of the observed effect in hierarchy as well as in the signs by the type of coupling is throughout observable in the individual branches for the other ALTERNATIVE PRINCIPLES (by the type of coupling) [23,177].

The first impression one gets is artificial, namely that the decision tree is not accompanied by an inverse effect in the "S"-branch compared to the "O"-branch. The stronger differentiating factors are not evident at increasing selectivity by compensation phenomena by this non-logarithmic scaling, because the selectivity is represented on a scale given in percentages. In the context with energy dependency of concentrations these aspects will be discussed below.

Nevertheless this shift from cooperativity to compensation can be demonstrated very nicely by pattern comparisons in the eliminations by selected examples, as we have already discussed in detail (Figs. 4.10 and 11) the exchange of one ALTERNATIVE PRINCIPLE. The patterns in Fig. 5.2 – concerning the selectivity of Δ1-olefine – are compared for *tertiary* amylchloride/-bromide/iodide with the alkylseries Me/Et/*i*-Pr/*t*-Bu in Na/K alcoholates under coherent conditions always in dimethoxiethane as the solvent. The tendency of alternatively increasing/decreasing in selectivity of Δ1-olefine is especially noteworthy in the iodide system with increasing steric requirements of the alkyls. This contradicts the explanations given in textbook (see data in Scheme 5.2).

According to an opinion strongly defended by H.C. Brown [178], the increasingly space filling bases are expected to interact less and less with the "sterically hindered" hydrogen at the secondary carbon atom, thus favouring a strengthened Δ1-olefine formation. The contradiction, that the Δ2-olefines are mostly favoured by using the "biggest" Na-alcoholates as bases and even the incomprehensable result, that the extent of control is reduced in the series (≈ 10%), can be explained with the help of the method of discontinuous variation [179] in the next chapter in context with the results in Fig. 6.10 [177].

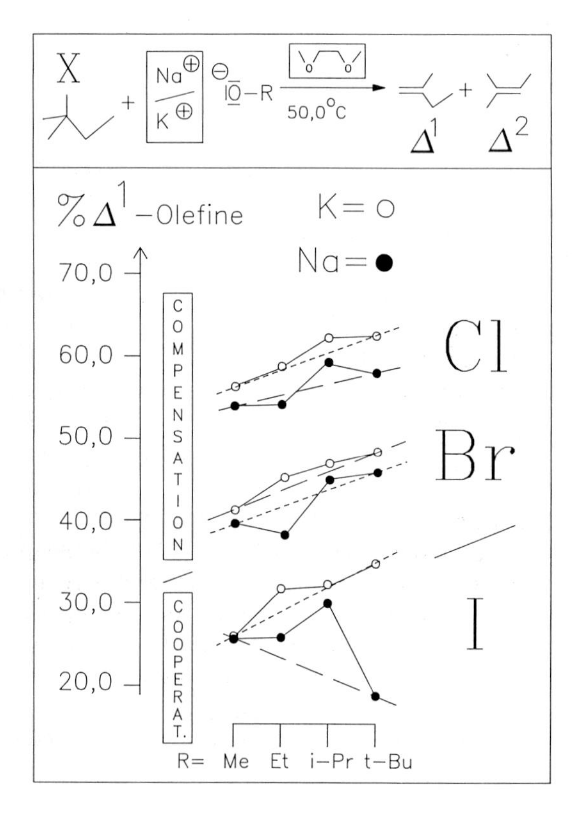

Fig. 5.2. Influence of alkyls in the alcoholates on regioselectivity by choosing Cl^-, Br^-/I^- and Na^+/K^+ as ordering factors

$$CH_3-CH_2-\underset{Br}{\overset{CH_3}{C}}-CH_3 \xrightarrow[-ROH,\ KBr]{+ROK} \begin{cases} CH_3-CH_2-\overset{CH_3}{C}=CH_2 & \Delta^1\text{-OLEFINE (HOFMANN)} \\ CH_3-CH=\overset{CH_3}{C}-CH_3 & \Delta^2\text{-OLEFINE (SAYTZEFF)} \end{cases}$$

BASE	CH_3-CH_2-OK	$CH_3-C(CH_3)_2-OK$	$CH_3-C(CH_3)(CH_2CH_3)-OK$	$CH_3-CH_2-C(CH_2CH_3)_2-OK$
Δ^1-OLEFINE %	38	73	78	89

Scheme 5.2

5.2 The Alternative Control of Knoevenagel Versus Michael Reaction of Mesityloxide

First of all we would like to describe those experiments, which lead in practice to Michael products exclusively or facilitate pathways especially to these products and influence considerably the reaction rates. Hereby we will slowly introduce the problems of alternative control to Michael or Knoevenagel products only by the type of catalysts concomitantly keeping constant the educts mesityloxide and the component with C-H-acidity – namely cyanoacetic acid methylester.

The conversion and the selectivity of the system described in Fig. 5.3a are totally insufficient with diethylamine as the catalyst, when the alkali-halides are not present as co-effectors. Only 5% of Michael product *M*, 3% of the Knoevenagel product *K* and 2% of "splitting products" *S* are formed from mesityloxide and cyanoacetic acid methylester after 72 h at 37 °C in 5 ml THF. The "splitting product" is formed from acetone (hydrolytic splitting of mesityloxide or Knoevenagel product *K*) and cyanoacetic acid methylester. While the potassium halides show practically no influence on product formation, the Michael reaction is dramatically activated – under simultanuous inhibition of the two other reactions – passing from sodium to lithium in the case of chlorides and bromides but applying the iodides passing from potassium to sodium. The system dependent change of Li/Na to Na/K as pairs of agonists/antagonists is accompanied by some parallel experience in quite different neural systems in front of and behind the brain barrier (for this see context of Fig. 9.4).

One could come to a hasty conclusion, that the solubilities of the salts in the chosen medium are decisive. However it is perhaps an unjustified prejudice to

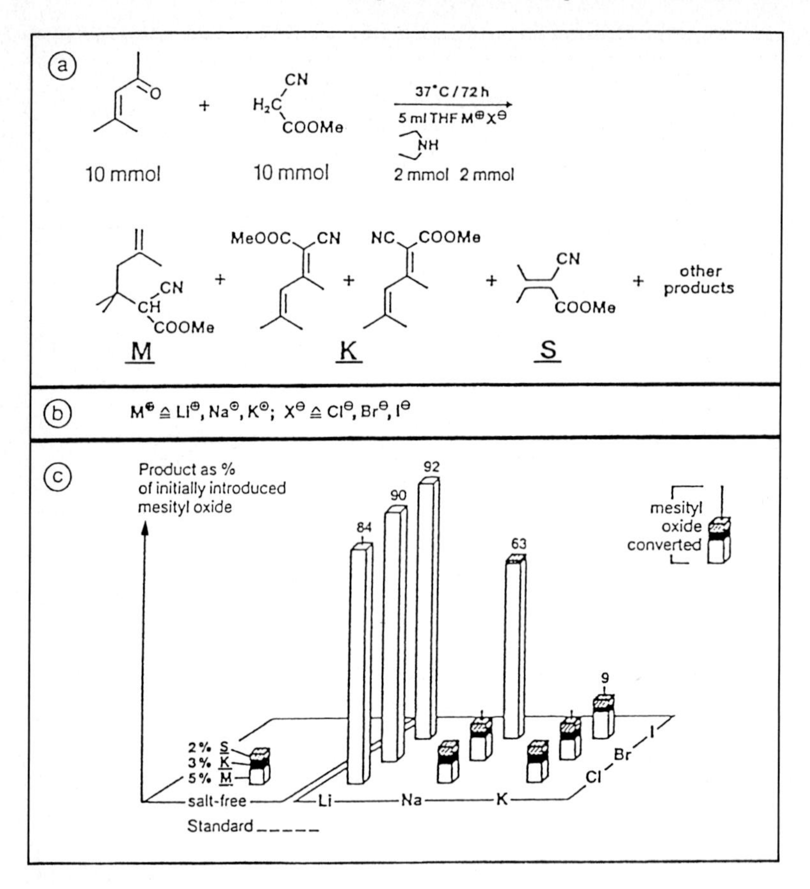

Fig. 5.3a. The influence of alkali-halides **(b)** on the unselective salt-free standard system **(c)** is investigated in a parent-system under coherent conditions

assume, that the transpassing of the phase solid/liquid alone causes effectiveness. This fixed assumption has to be relativated, when we consider the results of investigations done in comparison and presented in Fig. 5.4 by means of patterns [166]. The patterns of the influence of alkalichlorides on the reaction rates of the Michael products are practically inverted (alternative patterns of Li, Na, K-influences) in the "O/N" systems (for O/N variations and their effects see Figs. 2.15 to 18 and Figs. 9.3 and 4, too). The LiCl – being an effective activator in the "O" system – displays inhibition on the "N" system. The "O/N" systems differ by the –C≡N/–COOEt substituents and in addition by amine/enolate catalysts. Further preliminary experiments ensure, that the non-coherent conditions (20 °C, 40 h, 2 mmole catalyst respectively 37 °C, 72 h, 0,1 mmole catalyst) are not determining [180].

Very surprising controls of Michael/Knoevenagel reactivity (see Fig. 1.4) – perhaps understandable with the help of the CONCEPT OF ALTERNATIVE

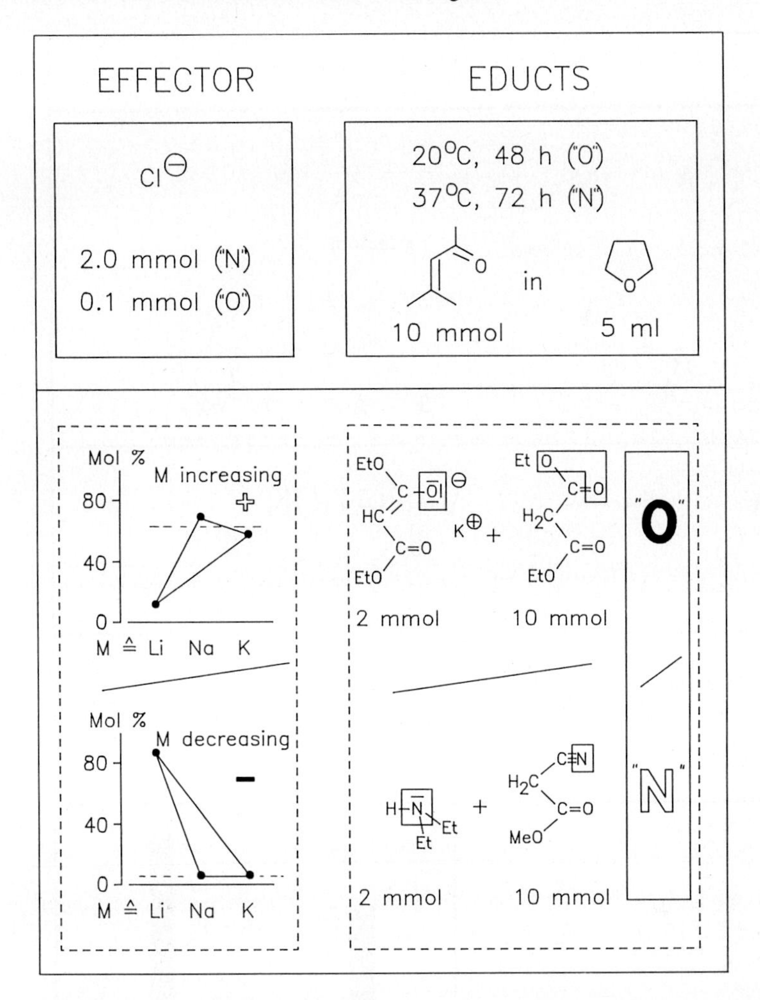

Fig. 5.4. A replacement of an "O"- by a "N"-system causes an inversion of reactivity patterns for $Li^+/Na^+/K^+$ in alkali chlorides as effectors. The results of the salt-free systems are shown by broken lines (Me/Et-variations in the estergroup have practically no influence)

PRINCIPLES – we would like to discuss by means of the results in Fig. 5.5. The knowledge extracted from the results and information in Fig. 3.20 and 21 was the starting point for this type of variation in the α,β-unsaturated oxo-compounds as demonstrated in Scheme 5.3. Thereby all H/Me substituents (alternative principles S/AS) should contribute an inverse influence mediated by the 2-/4-position in every case (an analogy to the inverse influence of the identical perturbation in 1-/2- respectively 1,4-/2,3-positions of 1,3-dienes: see Fig. 4.8, too). We have investigated the reactivity and selectivity of the four oxo-compounds shown in

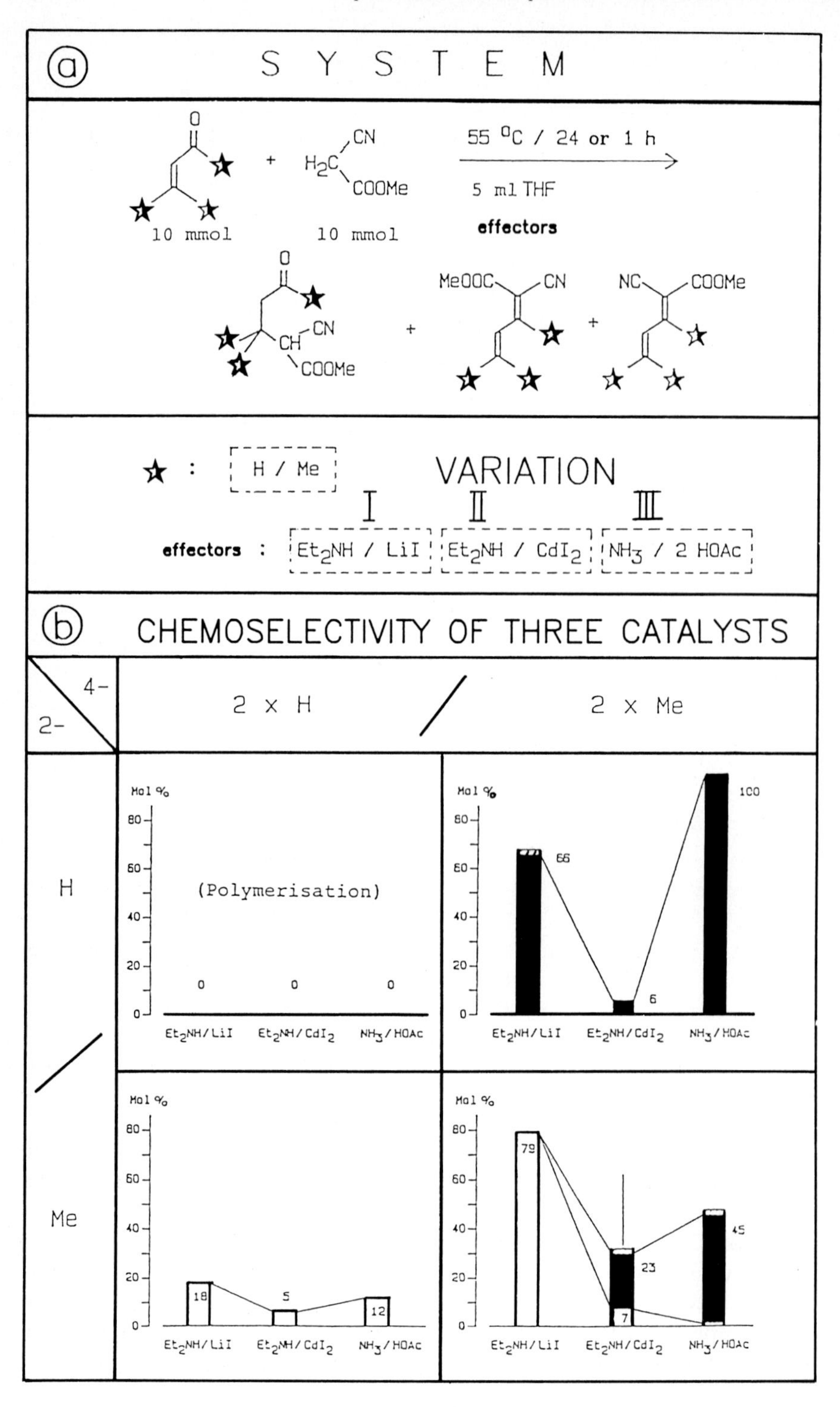

a
SYSTEM
55 °C / 24 or 1 h
5 ml THF
effectors
10 mmol
10 mmol
CN
COOMe
MeOOC
NC
H / Me
VARIATION
I
II
III
effectors :
Et2NH / LiI
Et2NH / CdI2
NH3 / 2 HOAc
b
CHEMOSELECTIVITY OF THREE CATALYSTS
4-
2-
2 x H
2 x Me
H
Me
Mol %
(Polymerisation)
Et2NH/LiI
Et2NH/CdI2
NH3/HOAc
66
6
100
18
5
12
79
7
23
45

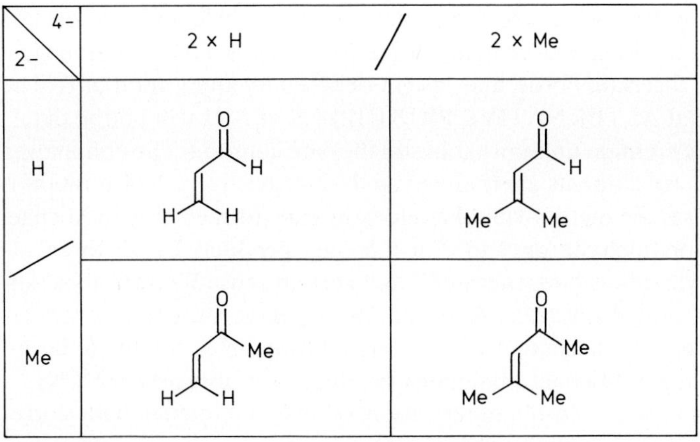

4- / 2-	2 × H	2 × Me
H	O, H, H, H	O, H, Me, Me
Me	O, Me, H, H	O, Me, Me, Me

Scheme 5.3

Scheme 5.3 with three different catalyst systems I to III under coherent conditions [166] (vide Fig. 5.5).

First we wish to account for the choice of the three catalyst mixtures I to III. The experiments for the evolution of the catalyst I applied in the Michael reaction are described in Figs. 5.3 to 4. We were less successful in our efforts, to catalyze the Knoevenagel reaction starting with identical educts. The, up to now, best catalyst mixture for this goal is II [166]. The high efficiency of ammonium-acetate as a catalyst for the Knoevenagel reaction is well known from textbooks. By optimizing this reaction we succeeded in proving, that the molar ratio 2:1 for acetic acid and NH_3 (a buffer) yields an optimum [166].

Within the CONCEPT OF ALTERNATIVE PRINCIPLES it would be interesting to investigate systematically, whether the alternative control of the Michael/Knoevenagel reaction (as 4q-/4q+2-system) may be induced starting from identical educts by the alternatives S/AS – located high in hierarchical order – by choosing an amine (a free π-electron pair) or an ammonium (no free π-electron pair available) alternatively as catalyst. A set of ALTERNATIVE PRINCIPLES in interacting subsystems must cause an inverse control in both cases. The question, whether the reactions of such alternative systems (vide Fig. 1.4) are submitted to a general phase relation control, is perhaps not set up in a general form, because e.g. Michael and Knoevenagel reactions and other alternatives are normally described in different chapters of the classical textbooks [181].

◄

Fig. 5.5a,b. Three different catalytic systems are investigated under coherent conditions in their efficiency and control on a parent system, in which by choice -H/-methyl group occupy alternatively 2-/4-positions of α,β-unsaturated oxo-compounds

These three catalysts were selected, because mesityloxide and cyanoacetic acid methylester could be transfered alternatively in Michael (I) or Knoevenagel products (III) with a certain borderline-case in between by application of II. The classes of individual ALTERNATIVE PRINCIPLES in I-III will not be determined in detail; they remain unknown, but still they are constant. The dominating influence of H/Me substituents alternatively in 4-/2-respectively 2-/4-position is surprising and causes the methyl-vinyl-ketone to be transformed only to Michael but 3-methyl-crotonaldehyde only to Knoevenagel products by all three catalysts. The mesityloxide is "degenerated" in a certain sense by methyl-substituents both in 2- and 4-position. Additionally we have observed a similar dominant influence of alternative H/Me substitution in a totally different reaction. Knoevenagel/Michael analogous products are observed with $Ni^{(0)}$-P-ligand-catalysts in a 2:1-co-oligomerization of two butadienes with correspondingly substituted oxo-compounds [127].

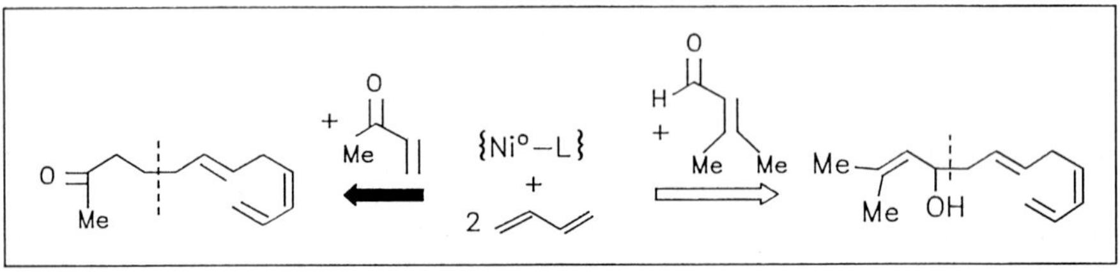

O Me + O Me {Ni°–L} + 2 H + O Me Me Me Me OH

The fact that acroleine is only polymerized by all three catalysts, can not be rationalized so far. But the normal classifications of Knoevenagel and Michael reactivity – e.g. by charge or orbital control – allow no interpretations without contradiction as it is demonstrated by the examples of molecular properties presented in Scheme 5.4 [182,183]

A parity correct interaction of all ALTERNATIVE PRINCIPLES by means of a general symmetry control for the whole reaction cascade could open up the way out of this dilemma.

5.3 Control of Asymmetric Synthesis in a Metal-induced Ketone Synthesis

A well known, carefully investigated system taken from the literature [184] is introduced by Fig. 5.6 to demonstrate the basic problem of the control phenomena to be discussed in this chapter. (a) Allylamines may be rearranged to enamines by Rh-catalysts in high selectivity. (b) Alternatively these Rh-catalyzed reactions may be utilized for the synthesis of *R*/*S*-enamines in high enantiomeric excesses either by starting from a stereochemically pure allylamine (e.g. *E*) by choosing paritetic ligands ((+)/(–)-BINAP, vide (c)) or by choosing alternatively

4- / 2-	LUMO COEFFICIENTS: 2 x H	LUMO COEFFICIENTS: 2 x Me	PARTIAL π-CHARGES: 2 x H	PARTIAL π-CHARGES: 2 x Me
H	+0,603 -0,497	+0,606 -0,527	+0,207 +0,038	+0,205 +0,161
Me	+0,616 -0,500	+0,619 -0,530	+0,260 +0,039	+0,257 +0,163

Scheme 5.4

one of the two allylamines (*Z* or *E*) and applying only one stereochemically pure ligand (e.g. (+)-BINAP, vide (d)).

While no enthalpy differences due to paritetically alternative ligands occur (energy degenerated alternatives of inverse parity) the additional differences in enthalpy have to be taken into consideration by applying *E*/*Z*-isomers. Nevertheless the parity – latently present in a subsystem (vide Fig. 10.1) – may be utilized for the control of alternative enantiomers, whenever the breaking of parity is accomplished in the system at least by one instance (here (+)-BINAP). We term this *parity*, being coupled with an additional difference in enthalpy, as com*p*lement*arity*.

We propose this, because *parity* is included in com*p*lement*arity* and com*p*lement*arity* contains more than *parity* alone (the enthalpy difference in addition: an amusing play upon words). On the other hand, we believe very strongly, that considerations of spatial alternatives alone do not allow a deeper understanding of asymmetric syntheses. The exclusive analysis of the "steric" situations at the re/si sides – alone space filling aspects – leads sooner or later to contradictions or even errors in logical typing (see e.g. Fig. 9.6 and context). Additionally we have to take into consideration the time by discussing electromagnetic interactions and strive to understand the self-organization of the systems to one or the alternative order by resonance phenomena of at least two alternative S/AS-interactions (vide Fig. 10.2).

A further hint to complementary ALTERNATIVE PRINCIPLES of other kind than *E*/*Z* is also given in the cited publication [184] (see [185,186] and many other publications). Figure 5.6e demonstrates graphically by means of a bifurcation which important information is lost in tables that one is usually faced with in many publications. The "achiral" four alkyls (Et/*i*-Pr) at the two phosphorus atoms of the applied chelating ligand with the constant (–)-DIOP-subsystem determine the corresponding (+)/(–) enantiomeric excess in the enamine. The very similar value in e.e. (38%/41%) is more or less accidental.

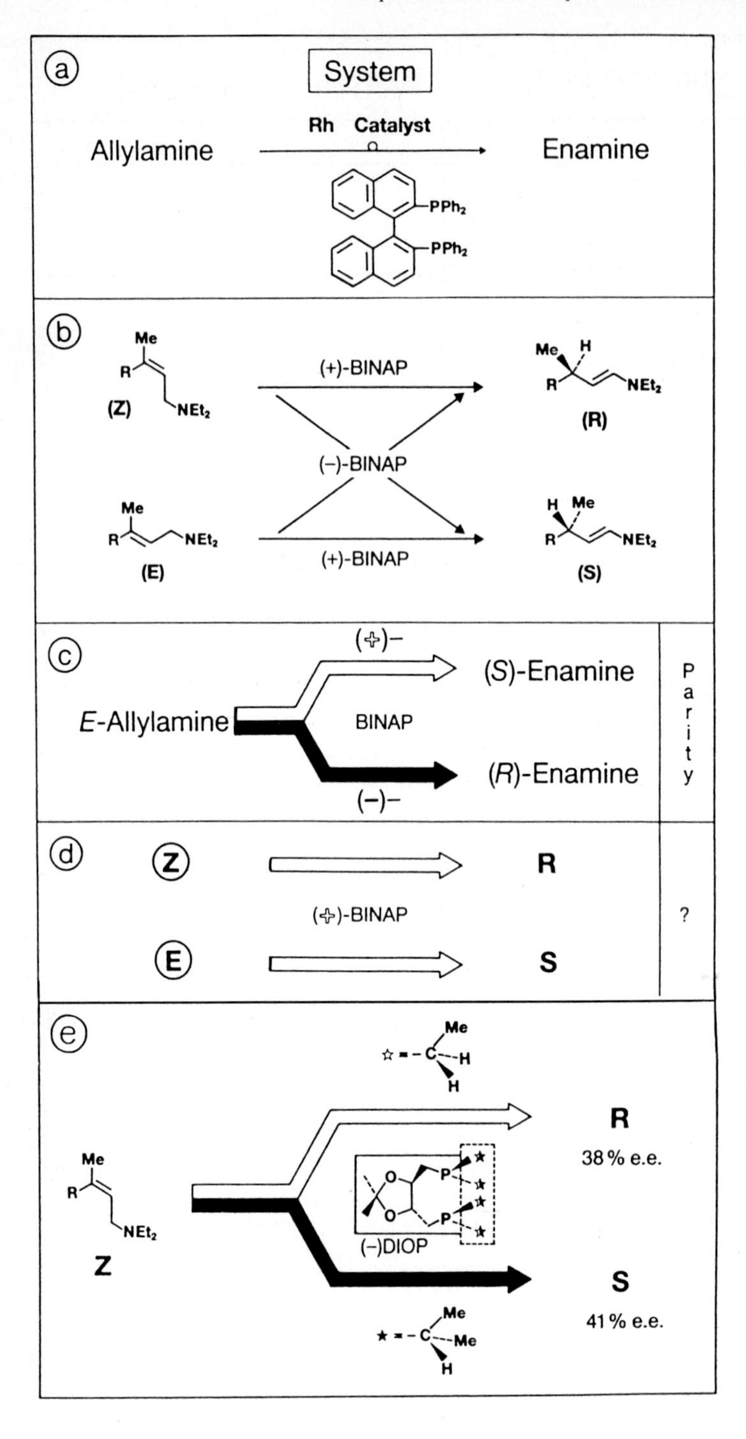
a
System
Allylamine
Rh Catalyst
Enamine
PPh2
PPh2
b
Me
R
(Z)
NEt2
(+)-BINAP
(−)-BINAP
(+)-BINAP
Me H
R
NEt2
(R)
Me
R
NEt2
(E)
H Me
R
NEt2
(S)
c
E-Allylamine
(+)−
BINAP
(−)−
(S)-Enamine
(R)-Enamine
Parity
d
Z
E
(+)-BINAP
R
S
?
e
Me
R
NEt2
Z
☆ = − C − H
Me
H
(−)DIOP
★ = − C − Me
Me
H
R
38 % e.e.
S
41 % e.e.

The difficulties in rationalizing, which may appear in the understanding of the control in asymmetric synthesis by application of complementary ALTERNATIVE PRINCIPLES, will be especially accentuated during discussion of the following figures. In all examples a maximum of 30% e.e. is obtained. Nevertheless we present these results, although the widespread, often single deciding factor of "belief" in quantity says that such results are not worth mentioning. Scientists, who are defeated by this heresy, have normally to accept, that they must discuss only "selected" results and have to forget most of their experimental investigations. In the following discussion it is of eminent importance to show *the way* leading to the target. The evolution of the pathways to pure enantiomers is more important than a high enantiomeric excess by chance. A further important aspect of the chosen procedure is the realization of both parities in the target compound even when starting the reaction – e.g. from the paritetically fixed "Pool of Nature" [36] – with preferentially only one of the two paritetic possibilities realized in educts or catalysts.

The system for this discussion is presented in Fig. 5.7. At first we will mention briefly all the many mechanistic studies in detail, which do not contribute at all to a real understanding of this asymmetric synthesis. The corresponding Ni-complex was intensively investigated to find out, how to avoid the *syn*/*anti*-isomerization of the methylgroups in 1- and 3-position in every case when adding ligands in different amounts of varying properties [187]. One consequence thereby is the stepwise addition of changing amounts of P-ligand at –20 °C to the dimeric 1,3-dimethyl-η^3-allyl-Ni-methyl. After exactly 30 minutes, the reaction mixture is cooled down to –78 °C and reacted with an excess of CO under normal pressure. The enthalpies of formation of the square-planar 1,3-dimethyl-allyl-Ni-L-complexes were as carefully investigated as the selectivity of C-C-coupling directly to 4-methyl-*E*-2-pentene or to 3-methyl-*E*-4-hexen-2-one – by additional insertion of CO – as a function both of type and amount of the ligands [62,187].

Knowing these conditions and results we investigated the optical induction in the ketone by applying P-ligands with substituents derived from (–)menthole. This procedure is facilitated by the fact that we were able to separate by condensation the above mentioned olefine which has no stereogenic center and the ketone of interest – the only volatile products – from the nonvolatile metal-complexes turning polarized light, too. Thus, controlled by internal standards, the percentages of the products are easily determined by GC. The enantiomeric excess can be calculated by simple measurement of the value of

Fig. 5.6a-e. The rearrangement of allyl-amines to enamines **(a)** can be controlled in the optical induction of the enamines starting from stereochemically pure E- or Z-allylamines applying alternatively (+)-/(–)-BINAP-ligands **(b)**. The two alternatives of control are individually presented once again in **(c)** and **(d)**. A third variation with complementary substituents is shown in **(e)**

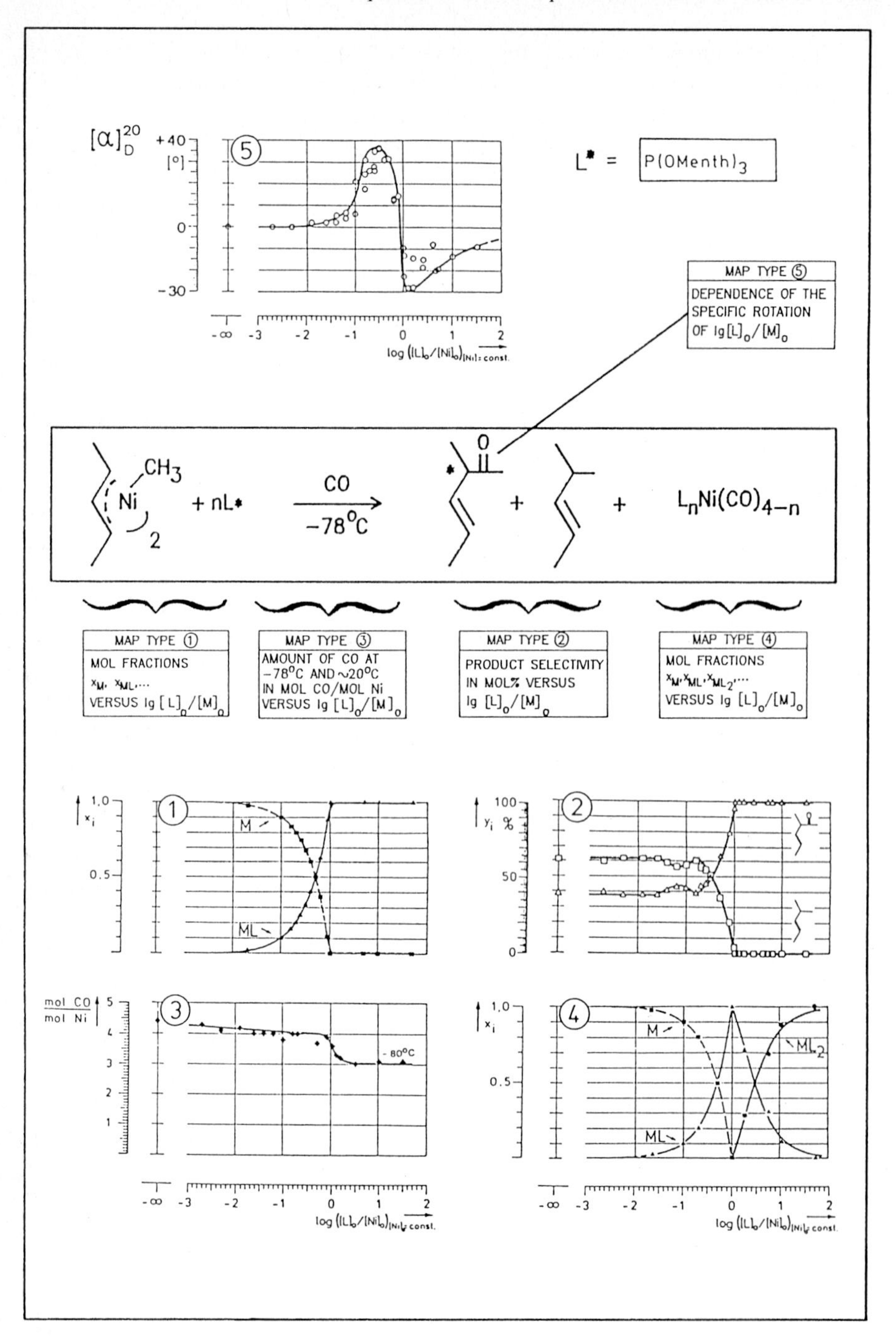

Fig. 5.7. A system for the metal-induced synthesis of optically active 3-methyl-*E*-4-hexen-2-one is presented together with the five possible investigations of system behavior by the stepwise variation of the amount of a ligand with parity breaking information

specific rotation. The (–)-menthyl subsystem was chosen as standard for parity breaking in the system, because, in this case, the enantiomeric (+)-menthyl group was available for us in larger quantities.

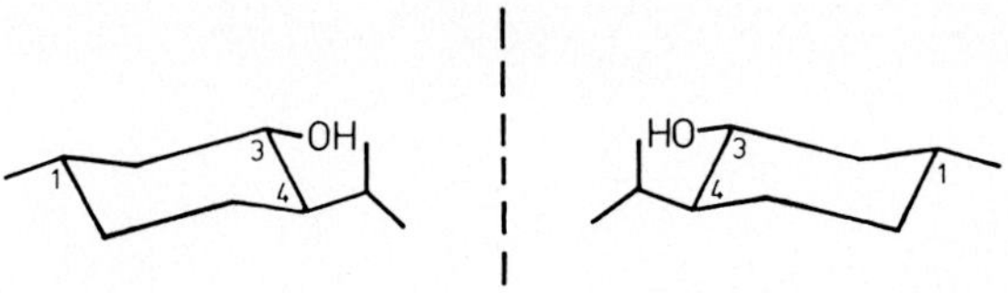

(1S,3S,4R)–(+)–Menthol (1R,3R,4S)–(–)–Menthol

Only those P-ligands were used in all experiments, in which the optically active groups (–)- or (+)-menthyl as well as the other substituents – isolated not optically active – were always introduced in a 2:1 or 1:2 ratio at the central atom, to avoid a new stereogenic center at the phosphorus. So-called Horner phosphorus compounds with three different substituents were not applied. The applied discontinuous method of INVERSE TITRATION will be discussed in Chap. 6. It can immediately be gathered from so-called "maps of rotational values" of type 5 in Fig. 5.7 (standardized on the formation of pure ketone (=100%), α_D^{20} = ±282±6 [24,70,188]) how type and amount of P-ligands influence the direction and extent of optical induction.

Further behavior of the systems are investigated by concentration control maps 1 to 4. Map of type 1: The degree of P-ligand association at –20°C after 30 minutes is determined by ^{31}P-NMR-spectroscopy. Using phosphites we recognized, that multifold associations might appear at super stoichiometric amounts causing the organic subsystems to be removed totally or in part from the complex by C-C-coupling before CO is added. Map of type 2: The selectivity of C-C-coupling (reducing elimination) and CO-insertion under coupling is determined. Discussion of the control of optical induction in the ketone will not be considered yet, because this problem may be investigated and solved separately [62]. Map of type 3: The consumption of CO correlates with the amount of ligand and may be important information about the selectivity of ketone synthesis in addition. In total, four coordination positions have to be occupied in the complex after reaction (xL + 4 – x CO). In addition one mole CO is consumed on complete conversion to the ketone [187]. Map of type 4: By this we were able to prove the kinetically caused distribution of L*-carbonyl-complexes. Their thermodynamic distribution is achieved after heating at 60°C in a closed system [19,187].

As presented in Scheme 5.5, we carried out NMR- and DSC-measurement at very low temperatures in addition which did not contribute essentially to any insight into the reaction mechanism or understanding of the direction and extent of optical induction.

The determination of the ratios in diastereomeric intermediates as a function of the amount and – at constant Ni:L ratios – of the type of applied, optically active P-ligands (at 20°C), e.g. gave no correlation at all (completely random) to the observed enantiomeric excesses in the ketone [70].

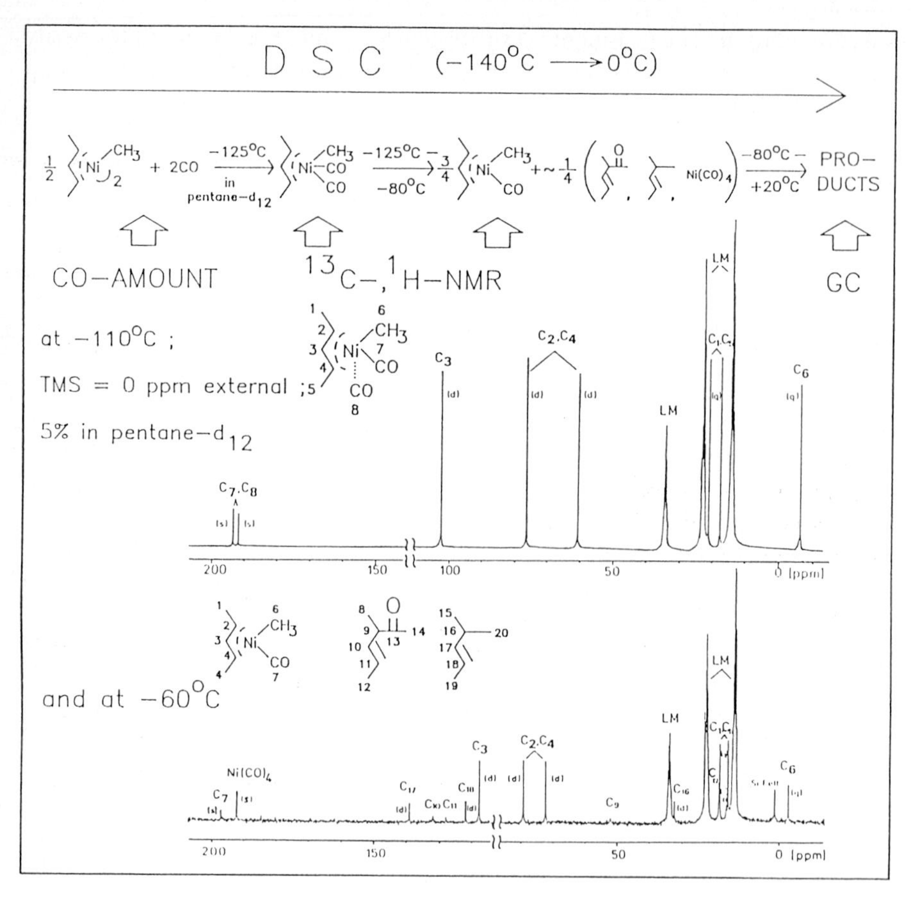

Scheme 5.5

Therefore we tried to get qualitative rationalization by the CONCEPT OF ALTERNATIVE PRINCIPLES resorting to INPUT/OUTPUT relations (for the external dynamics: see Chap. 7). The most important insights will be presented by the examples in Figs. 5.8 to 10.

As described in the literature (e.g. [184]) and exemplified in Fig. 5.6e, the type of – isolated – not optically active substituents may determine the direction of optical induction. Three further examples of this phenomenon at the Ni-system are presented in Fig. 5.8. They are selected to demonstrate clearly possible errors in logical typing. In (a) the direction and extent of optical induction in the ketone

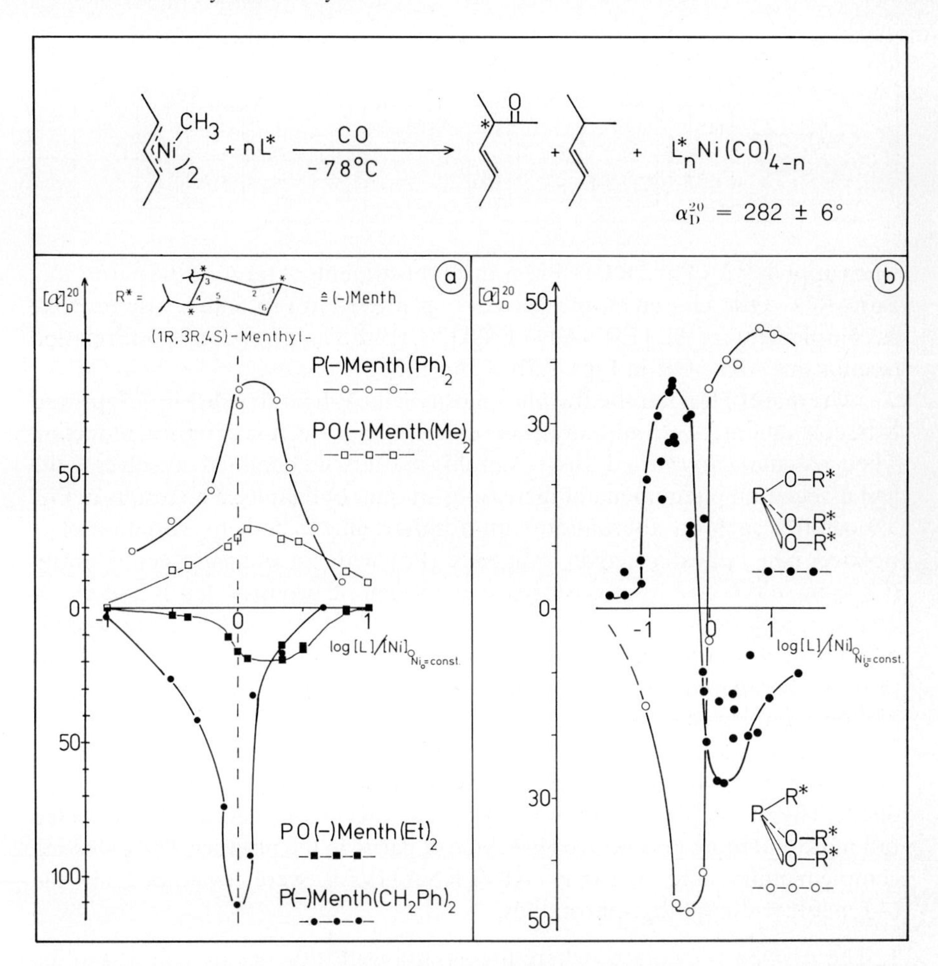

Fig. 5.8. The alternative direction of optical induction in the ketone of the presented system is effected by replacement of 2×Me- by 2×Et- respectively of 2×Ph- by 2×$PhCH_2$-groups in a P-ligand with alternatively (−)-menthyl/O(−)-menthyl subsystems. In addition the inverse influence of -R*/-OR* substitution is demonstrated

synthesis is fixed by the option for 2×methyl-/2×ethyl- or 2×phenyl-/2×benzyl substituents at the phosphorus. The reproducibility was controlled by applying the corresponding (+)-menthyl derivatives. All experimental curves (not represented here) showed the patterns of mirror images [154].

The result should not lead to the false conclusion that substituents like methyl and phenyl respectively ethyl and benzyl behave similarly in determining the parity of the enantiomere in excess.

It has to be considered necessarily, that the O(−)-menthyl group (−OR*), when applying the alternatives 2×Me/2×Et, and the (−)-menthyl group (−R*),

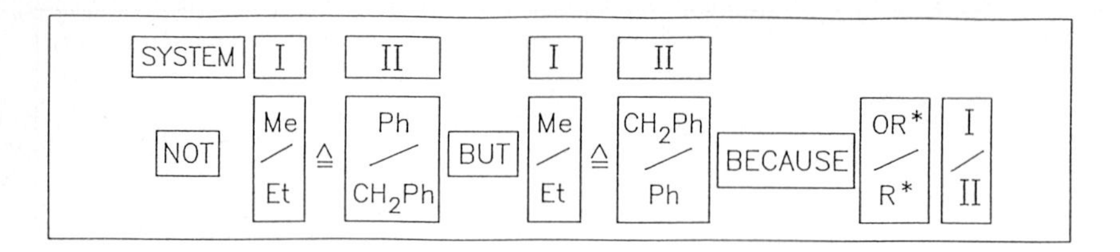

when applying 2 × Ph/2 × CH_2Ph, as third substituents determine the parity, too. But –R*/–OR* chosen as alternatives – as proved in Fig. 5.8b – also provoke as complementary ALTERNATIVE PRINCIPLES an inversion of information (see for this –R/–OR in Figs. 2.7b, 8 and 13b).

The more or less tetrahedral phosphorus in the P-ligand (vide Fig. 2.3) proved to be convenient for examining an alternation in the direction of optical induction when systematically varied. Just to simplify matters, we confined ourselves to the first association phenomena of increasing amount of ligands. The results in Fig. 5.9 confirm such an alternation with absolute effects both by variation of – isolated non optically active – ligands (Ph) with an optically active group (O(–)-menthyl) and by choosing two enantiomeric groups (O(+)- and O(–)-menthyl) for this procedure. In (b) the curves (1) and (4) respectively (2) and (3) exhibit mirror patterns. The curve (4) – investigated at the beginning of this project – compared to curve (1) demonstrated to us that coherency of conditions had been improved constantly.

The systematic evolution of a system proves to be a preparatively laborious procedure, especially when starting from a situation faced with a low percentage of e.e.. Therefore, Fig. 5.10 shows only the early levels of a decision tree directed to a more and more pronounced breaking of parity in the product. The following complementary and paritetic ALTERNATIVES were examined in the (–)-menthyl-dinaphthyl-phosphites.

1. The *position* is changed, where the naphthylsubstituents are coupled to the phosphorus of the ligand via the oxygen.
2. The *structure* may be changed from OPEN to CYCLIC (vide Fig. 4.16) by the direct unifying of two naphthylgroups in *ortho*-positions. In the case of 2-naphthyl-derivatives (β-position) this "cyclization" is realized in the α-positions (vide Fig. 5.10b).
3. The diastereomeric mixture (atropisomeric *R*/*S*-subsystems formed in a 1:1 ratio in synthesis can be separated by repeated crystallization.

Fig. 5.10 demonstrates the results of the experiments. In (a) the values of the maxima derived from "rotational value maps" at the first association phenomenon are listed on a scale, from which a threefold dual decision tree is constructed. The individual effects of the realized variations are depicted in Fig. 5.10c to e [154]. The example of bifurcation resulting in higher and lower optical yields of the β, γ-unsaturated ketone in this special case underlines the quite general statement, that enantiomeric (paritetic) and even complementary conformations contribute decisively to the cancellation of the maximally possible induction. Here the 1:1 mixture of conformations in binaphthyl derivatives may

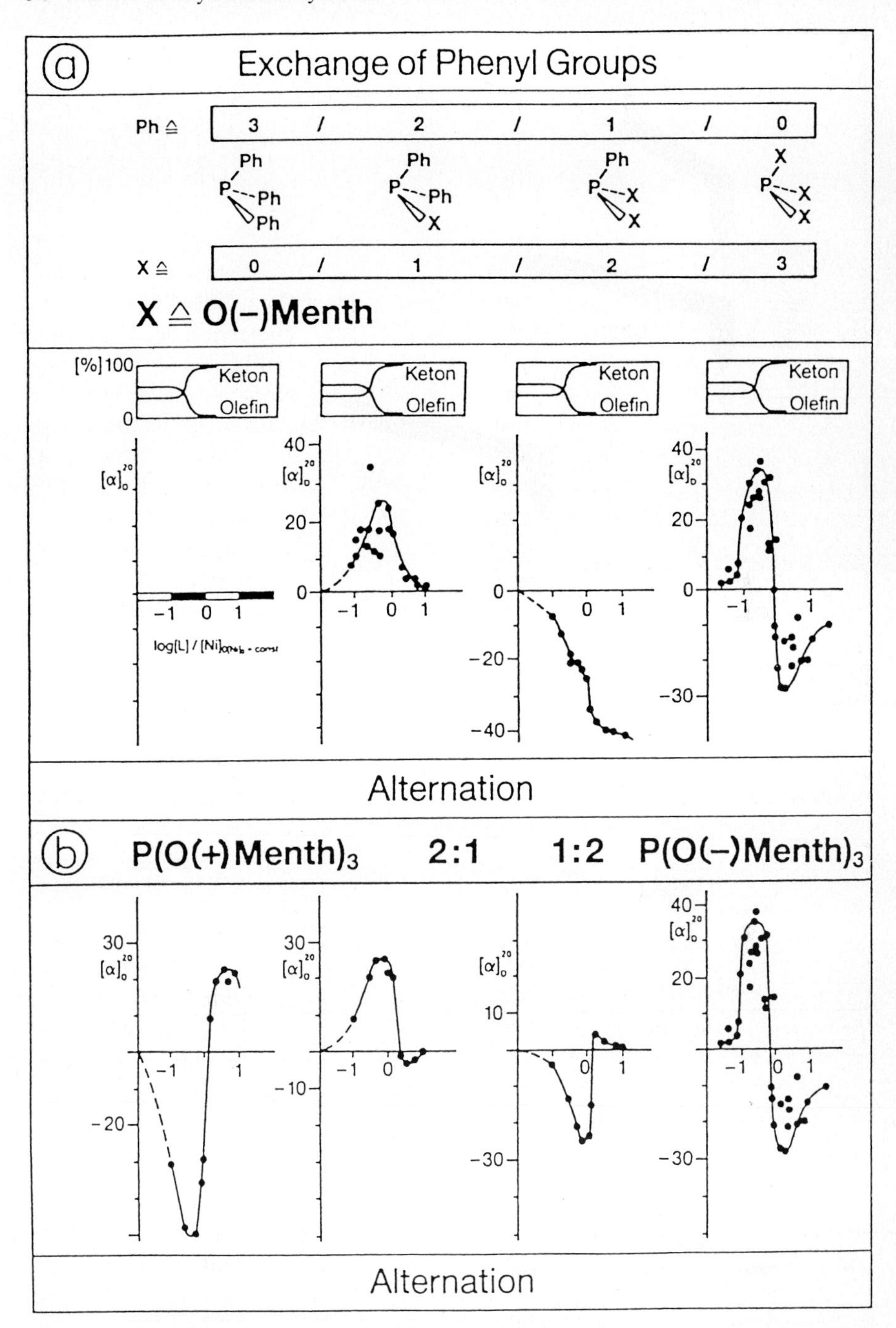

Fig. 5.9a,b. Alternating phenomena of optical induction in the ketone of the system presented in Fig. 5.7 and 8 are observed, when **(a)** starting from triphenylphosphane or **(b)** from P(O(+)-menthyl)$_3$ the three substituents present are successively replaced by -O(–)-menthyl groups

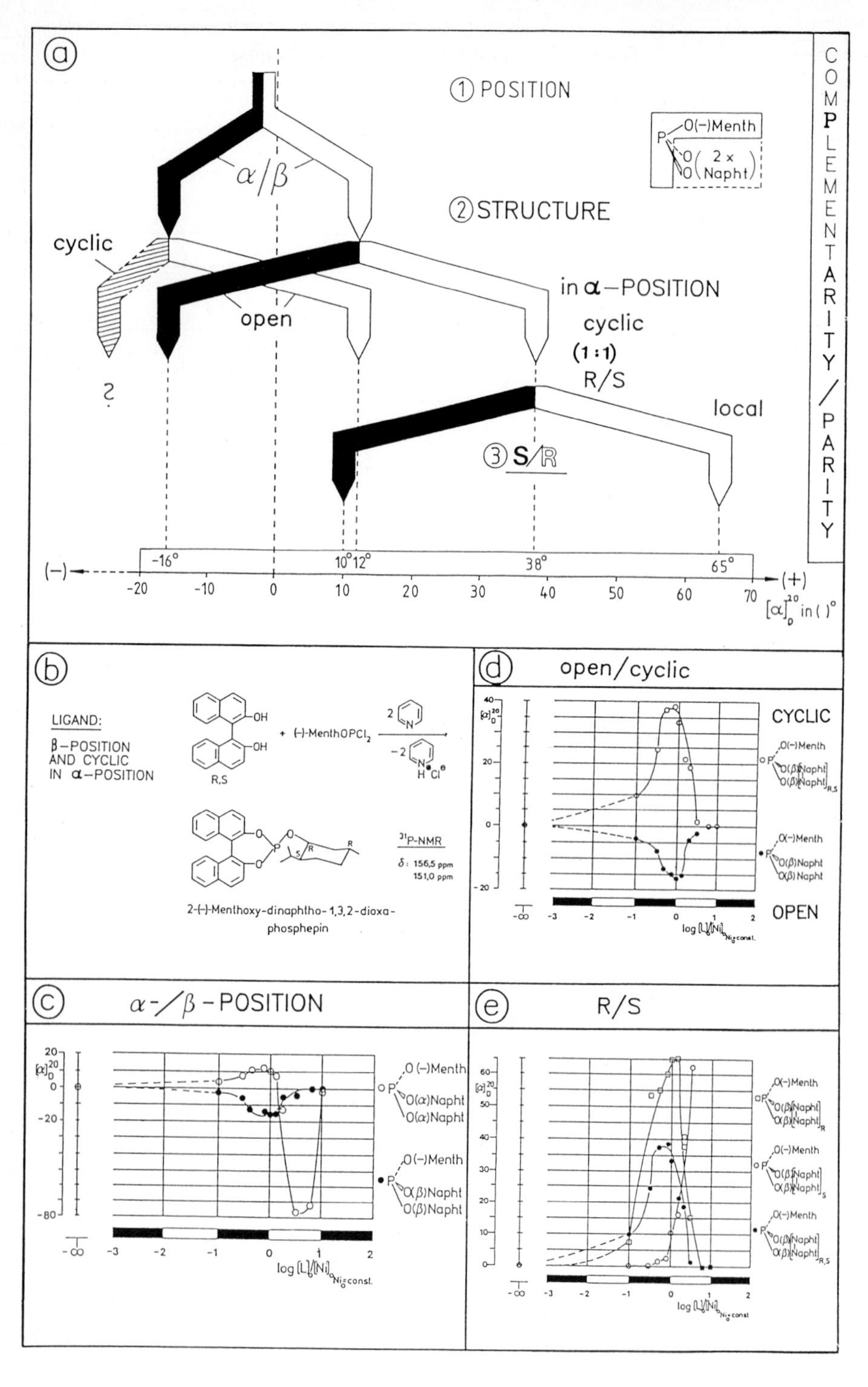
a
① POSITION
α/β
② STRUCTURE
cyclic
open
in α–POSITION
cyclic
(1 : 1)
R/S
local
③ S/R
-16°
10° 12°
38°
65°
(−)
(+)
$[\alpha]_D^{20}$ in ( )°
COMPLEMENTARITY / PARITY
b
LIGAND:
β–POSITION AND CYCLIC IN α–POSITION
R,S
+ (−)-MenthOPCl₂
^{31}P-NMR
δ: 156,5 ppm 151,0 ppm
2-(−)-Menthoxy-dinaphtho-1,3,2-dioxa-phosphepin
d
open/cyclic
CYCLIC
OPEN
c
α-/β-POSITION
O(−)Menth
O(α)Napht
O(β)Napht
e
R/S
$\log [L]_0/[Ni]_0$ Ni_0 = const.

be replaced by the individual conformers after separation, which is possible because of the high activation energy for inversion of local order. The distance-rule, assuming a misleading correlation between the distance of the reaction center and the stereogenic atom and the extent of optical yield, has to be replaced by a compensation rule, which describes a correlation of optical yield with the minimization of possible molecular arrangements of alternative states *and* the cooperativity or compensation of activation/inhibition of alternative processes.

Masamune [170] proposed for the double diastereo-differentiation – being a special case of complementary ALTERNATIVE PRINCIPLES – the terms "matched/mismatched pairs". The effect is based on the electromagnetic interactions of all states with paritetic or complementary ALTERNATIVES, which cooperate to a maximum or minimum (related to one parity) or – by two types – compensate (for the difference between parity and complementary, vide Fig. 4.1). We will avoid the term "chirality" and related terms in this chapter and the reasons will be discussed in connection with Figs. 9.5 and 6. The two hands of a living being are complementary, but not paritetic. A further optimization of the above mentioned asymmetric synthesis may be achieved by variations presented in Scheme 5.6. It would be possible to achieve the synthesis of the enantiomeric ketone e.g. by changing -R*/-OR*, if we succeeded in the formation of one enantiomer by good cooperativity of all the ALTERNATIVE PRINCIPLES.

5.4 Influence on Structures and Processes in Transition Metal Complexes

The cooperation of only two pairs of ALTERNATIVE PRINCIPLES will be discussed with respect to structures (for structures and processes vide [32,39]) in the dimeric allyl-Ni-X compounds. As demonstrated in Fig. 5.11a the spatial arrangements in the dimeric allyl-Ni-O-/S-R complexes is determined by the heteroatoms O/S in the bridgehead position. While in the "O"-system a square planar arrangement in the four membered ring with 2 × O- and 2 × Ni-atoms can be observed by X-ray analysis accompanied by a quasi chair arrangement of the two allylic groups, one can identify in the "S"-system, a tilted four-membered ring with a "boat-like" arrangement of the allylic groups [105]. The alternative

Fig. 5.10a. Complementary ALTERNATIVES of position (α-/β-) and of structure (OPEN/CYCLIC), see **(b)**, as well as local parity (*R*/*S*) in BINAP-phosphites of the system presented in Fig. 5.7 to 9 determine the form of a multi-dual decision tree for the maxima of optical induction of the first association phenomena in an α,β-unsaturated ketone. The "rotational value maps" of the individual pairs of ALTERNATIVE PRINCIPLES are presented in **(c)**, **(d)** and **(e)**

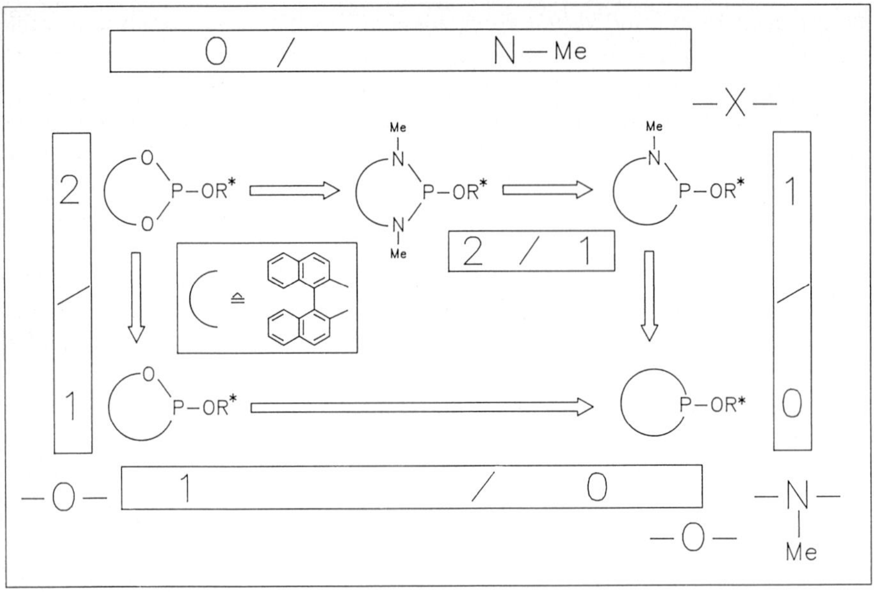

Scheme 5.6

arrangement of the alkyls at the sulfur in the following two complexes is of interest, too.

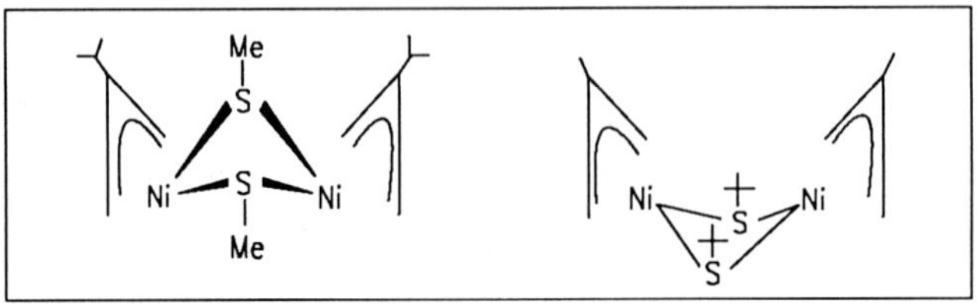

It is still an open question, whether the two variations (Me/*i*-Pr in 2-position of the allylic groups or – more realistic – Me/*t*-Bu at the sulfur atoms) are higher in hierarchy. The systems are not comparable at all.

But intensive, temperature dependent NMR-investigations here not discussed in detail, prove the trend conclusively, namely that the replacement of methyl by tertiary butyl groups at the bridge-head O/S-atoms influences the alternative spatial arrangements of the four-membered rings inversely (planar/tilted). This is schematically represented in Fig. 5.11a.

A branch of a digital decision tree, derived from a specific reaction of certain subsystems in complexes by Mingos, Davies and Green [189] is printed in Fig. 5.11b as a first general application of hierarchical ordered decisions of ALTERNATIVE PRINCIPLES. The system and the application of the rules will not be presented here in more detail, because this is done by numerous examples in a review [190].

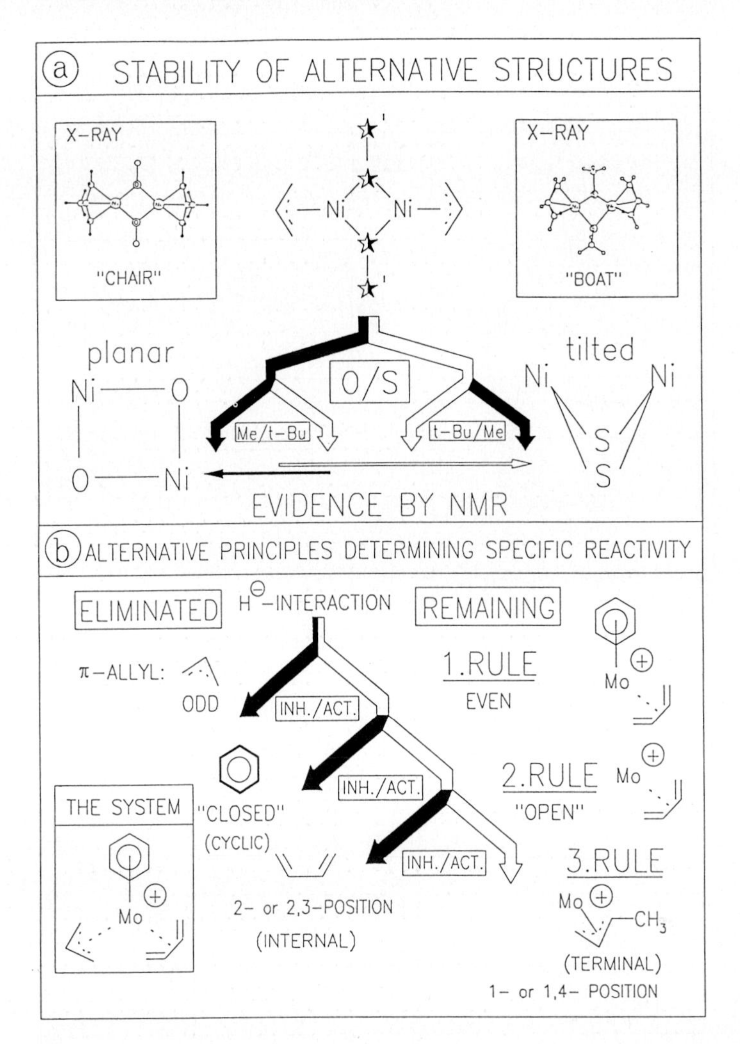

Fig. 5.11a. Cooperativity of -OMe respectively -SMe leading to alternatives of structures in four-membered rings (PLANAR/TILTED) and compensating influence of t-Bu-groups at the bridge-head positions in O/S-dimers. **(b)** The three rules for the importance of hierarchical order of ALTERNATIVE PRINCIPLES in the nucleophilic substitution of π-systems of positively charged transition metal complexes in form of a branch of a multi-dual decision tree

One remark seems to be of special importance. Whenever the alternative decisions of preferred reactivity by the proposed rules are not unequivocal, it is possible to "readjust" the precision of the rules according to the CONCEPT OF ALTERNATIVE PRINCIPLES with the help of systematic variations in any subsystem (reagents, effectors, solvents and so on), which is in interaction with the complex in the cascade of steps.

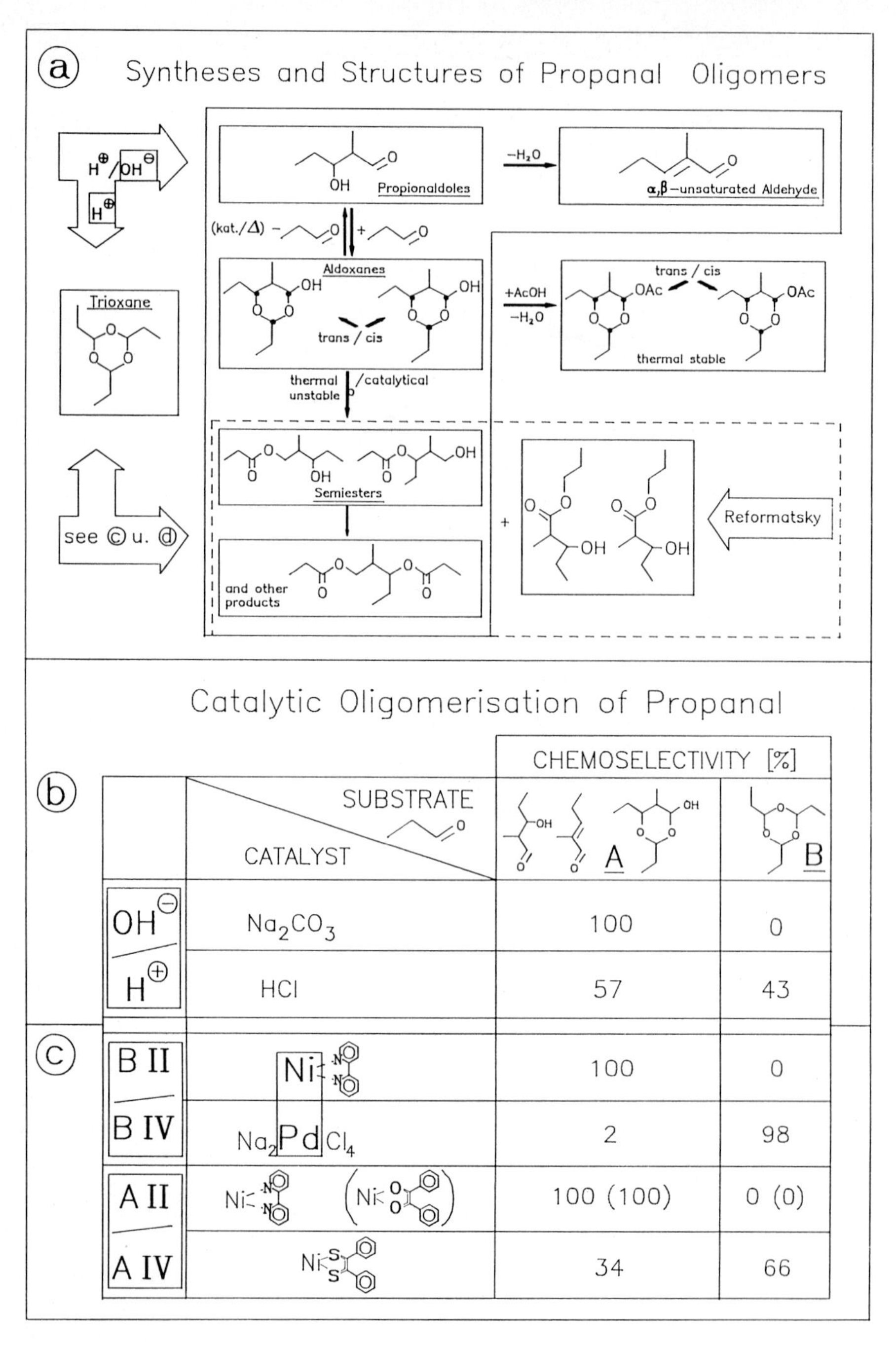

Fig. 5.12a. Complex product spectrum of propanal by the H^+/OH^--catalyzed **(b)** or Ni/Pd-catalyzed **(c)** oligomerization. **(d)** Abstract presentation in the form of bifurcations (see also scheme 41 in [22])

5.5 A Comparison of the Catalytic Oligomerizations of Propanal

Finally we would like to present an example, at present not widely applied, that may in the future have even greater potential in specific control of organic synthesis than the tremendously increasing use of metalorganic reagents: namely the oligomerization of heteroolefines. The co-oligomerization of hetero-π-system with carbon-π-systems has been very intensively investigated, too (see e.g. [191,192] and Fig. 3.19c). All available methods (see Fig. 8.6) should be applied using the CONCEPT OF ALTERNATIVE PRINCIPLES (vide Figures 8.1 and 2) to provide solutions to the problem of unsatisfactory turnover numbers [127,193].

The oligomerization of propanal leads to a great range of product mixtures such as isomeric α,β-unsaturated aldehydes, aldoles, aldoxanes and so on (vide Fig. 5.12a). The analysis of such product mixtures is made considerably more difficult by the equilibria between aldoles and propanal (or aldoles and oligomers) with aldoxanes, which are thermically and catalytically rearranged to the semiesters shown. In addition water may be eliminated from aldoles by heat and catalysts. The analytical problems of these mixtures could be solved satisfactory [195] only by application of special GC-techniques with direct inlet [194].

Only the control of the collective pathways to the product mixtures A and B will be discussed omitting much unnecessary detail. Comparing the data in Figs. 5.12a and b it is remarkable, that the original unsatisfactory selectivity to products of type B by H^+-catalysis can be overcome by application of transition metal catalysts (Na_2PdCl_4). Interestingly the reaction pathways to A and B can be controlled both by alternative metals (Ni/Pd) and by alternative heteroatoms as central atoms of chelating ligands at Ni-catalysts. Hopefully the reader is becoming more and more familiar with the abstract comprehensive and comparative presentation as shown in Fig. 5.12d [40].

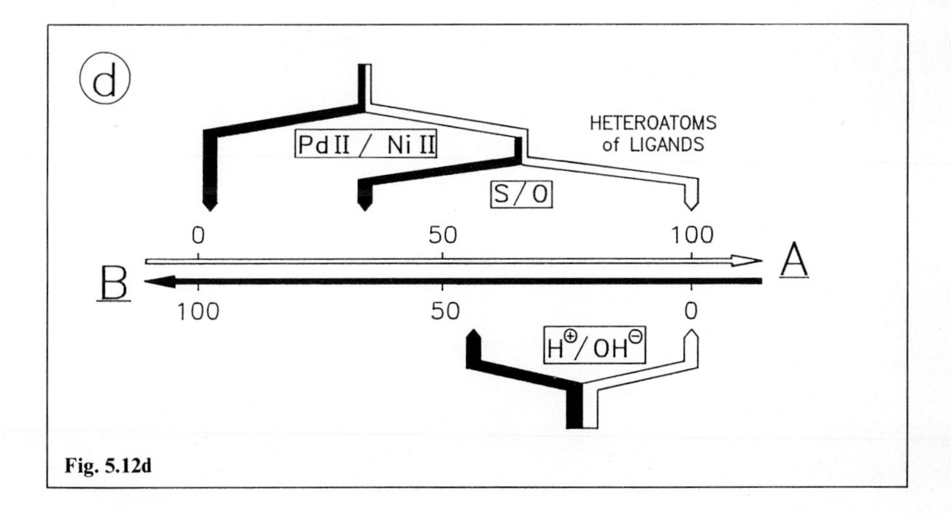

Fig. 5.12d

A comprehensive review of Ni/Pd-catalysts in the oligomerizations and co-oligomerizations of strained olefines has been published by Binger [196].

With the broad applications of the CONCEPT OF ALTERNATIVE PRINCIPLES – sometimes presented in detail and sometimes only by suggestion – for different chemical systems the scope and limitations of this procedure can only be sketched out for the experimental evolution of more complex systems. At the moment, we are not aware of any great barriers. Some new aspects will be introduced in the next chapters.

6 The Discontinuous Method of INVERSE TITRATION

"Any complex system of matter is completely characterized by the spatial and temporal distribution of its various individual constituents, expressed in terms of their population densities."

The Art of Titration [47]
M. Eigen and R. Winkler-Oswatitsch

As previously mentioned, added catalysts, effectors or solvents may crucially improve the selectivities and reaction rates of chemical processes. Therefore as we once said provocatively: "A lot of help by amount and property of compounds helps a lot!" [18]. The crucial question remains: "What kind of help by property and how much of the compound?" The first part of the question "What kind of help" we tried to answer with the CONCEPT OF ALTERNATIVE PRINCIPLES. In this chapter we would like to approach the other problem "how much" in more detail.

It is accepted as a truism, that the amount of a compound itself controls the behavior of a chemical system. In Fig. 6.1 it is shown by two extreme examples, that catalytic or stoichiometric amounts of a similar or even the same compound may give totally different orders in the product spectrum. Aliphatic aldehydes in high concentration relative to the ylide are transformed in (a) into oligomeres of these aldehydes (vide Fig. 5.12). Higher associations than 1:1 ratio (Wittig reaction) are built up by the high aldehyde concentration and catalysis occurs [32].

6.1 Evidence in Support of Concentration Effects: A Summary

Following the quotation taken from the publication "The Art of Titration" [47] at the beginning of the chapter, one of the most urgent problems in chemical processes is to find out quantitatively the effects of the necessary amounts of all participating components in a complex chemical system by a general procedure. This is especially true for unevolved systems.

Totally different titration techniques can be applied, which amongst others are described in the following textbook [46] and in the article mentioned above [47]. Two main reasons of overwhelming interest exist for titration:

1. The unknown amount of a constituent in a chemical system has to be determined analytically.
2. Various information about and determination of equilibria (e.g. K) or steady state parameters (e.g. K_M and so on) is to be obtained for problems in synthesis.

If, for instance, the association phenomena are determined spectroscopically as a function of relative added concentration and plotted logarithmically, curves

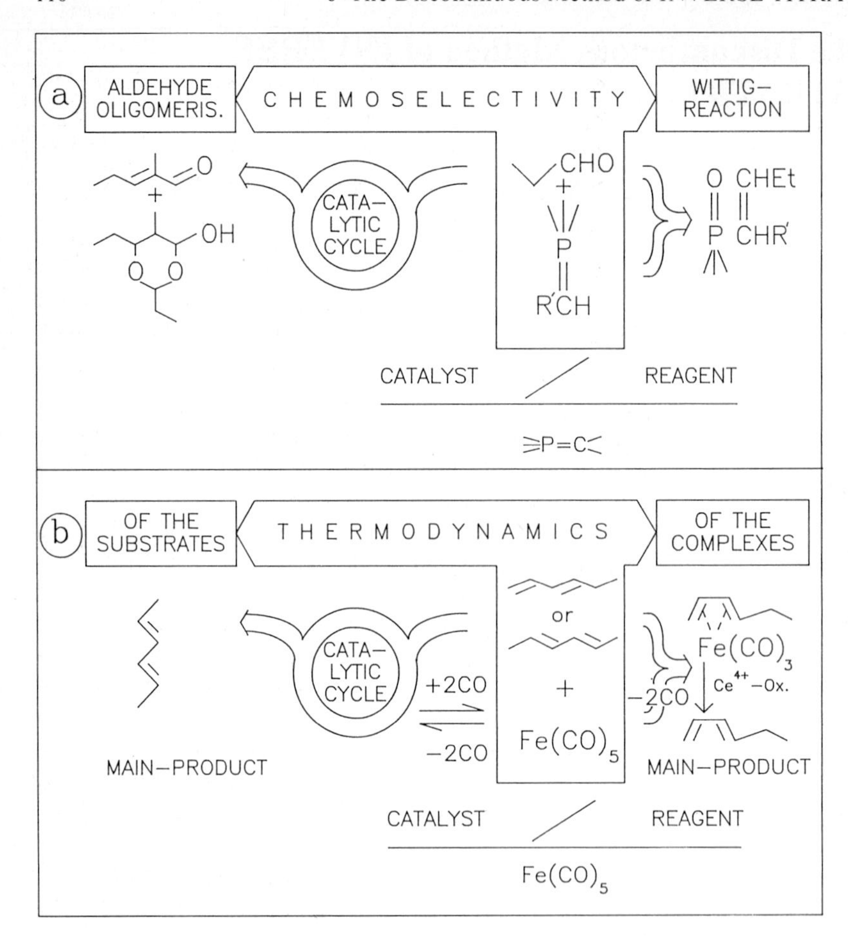

Fig. 6.1a-b. Comparison of two different systems by application of stoichiometric (right hand side) or catalytic amounts of effectors or educts. **(b)** [197]

with different shapes may occur as shown in Scheme 6.1. Three types of curves may be distinguished:

Type I: The so called classical titration based upon a well defined determination of endpoint ($K ---> \infty$), by which any added amount of a compound is consumed by total formation of an associate (the thicker line in Scheme 6.1 is defined as a "stoichiometric curve"). By kinetic separation, if only parts of one constituent are available for association, curves placed on the left hand side of this "stoichiometric line" can be observed.

Type II: The totally symmetrical curves, from which one may easily obtain the mass law parameters from the extreme values; they are well known to every research-chemist from titrations in aqueous solutions [46].

Type III: The distorted curves in between these two extreme forms (in this area the dynamic titration techniques are applied with great success [47]).

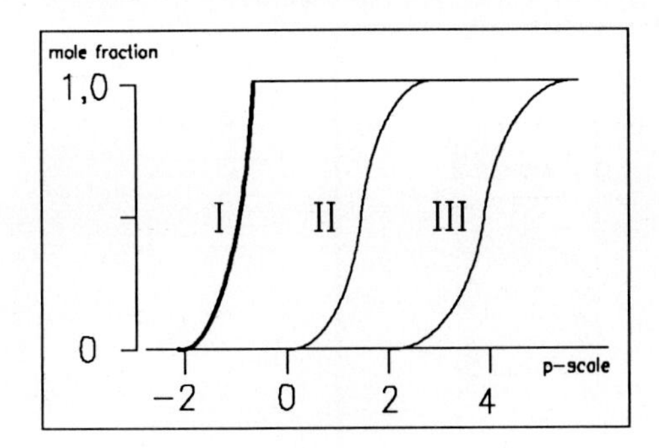

Scheme 6.1

The real problem concerning "The Art of Titration" is to choose the right technique for the problem under investigation. In unevolved metal-induced and -catalyzed processes we normally will not determine the unknown amount of a component in the system by adding a well defined amount of a standard solution (normal titration). Moreover we would like to know in a given standard solution, e.g. of a catalyst-metal in an educt or educt mixture, the still unknown quantity of one or more ligands and co-effectors for producing an optimal change in the behavior of the overall system to give the desired process. That means exchanging the independent and dependent values within our procedures step by step. Therefore we call this technique "the discontinuous method of INVERSE TITRATION" (vide Fig. 6.2a) with examples for the influence on patterns in thermodynamic equilibria in both types of plotting in (b) (for similar investigations of metalorganic systems in the literature see e.g. [198,199]). All associations in the whole system will be dealt with by this method, it will be "titrated" step by step.

In order to analyze catalytical systems by the discontinuous method of INVERSE TITRATION – an individual step by step process – one has to investigate at least three to four individual catalytic reactions for each chosen power of a tenth of the external ligand-to-metal ratio. To obtain this information efficiently with respect to time needed and the amount of chemicals, we carried out a special experimental procedure e.g. on the 1-ml scale (for experimental details see [39,63,64,200]: a flow chart is given in Fig. 6.3).

The procedures are different in detail from system to system. To describe one of them: Standard solutions of the controlling ligand and of the used metal complexes are prepared under an inert gas atmosphere, both including internal standards, e.g. tetralin and n-dodecane, to determine by gas chromatography the ratio log ($[L]_0/[Ni]_0$) from the ratio of these two components. The standard solution of the ligand is then diluted in steps of powers of ten.

Using the apparatus described in Fig. 3.2.1 of Ref. [200] the required quantities of the ligand and the metal solution are placed in glass capillaries (Duran 50, $\phi = 8.0$ mm, $l = 27$ cm) and filled with solvent and substrate up to 1 ml. As an example, we used for the investigations of the catalytic systems nickel/ligand/butadiene the ratio 1:(10^{-6} up to $10^{1.5}$):200. The nickel concentration ($[Ni]_0$) was constant for the whole series of experiments. The tubes were

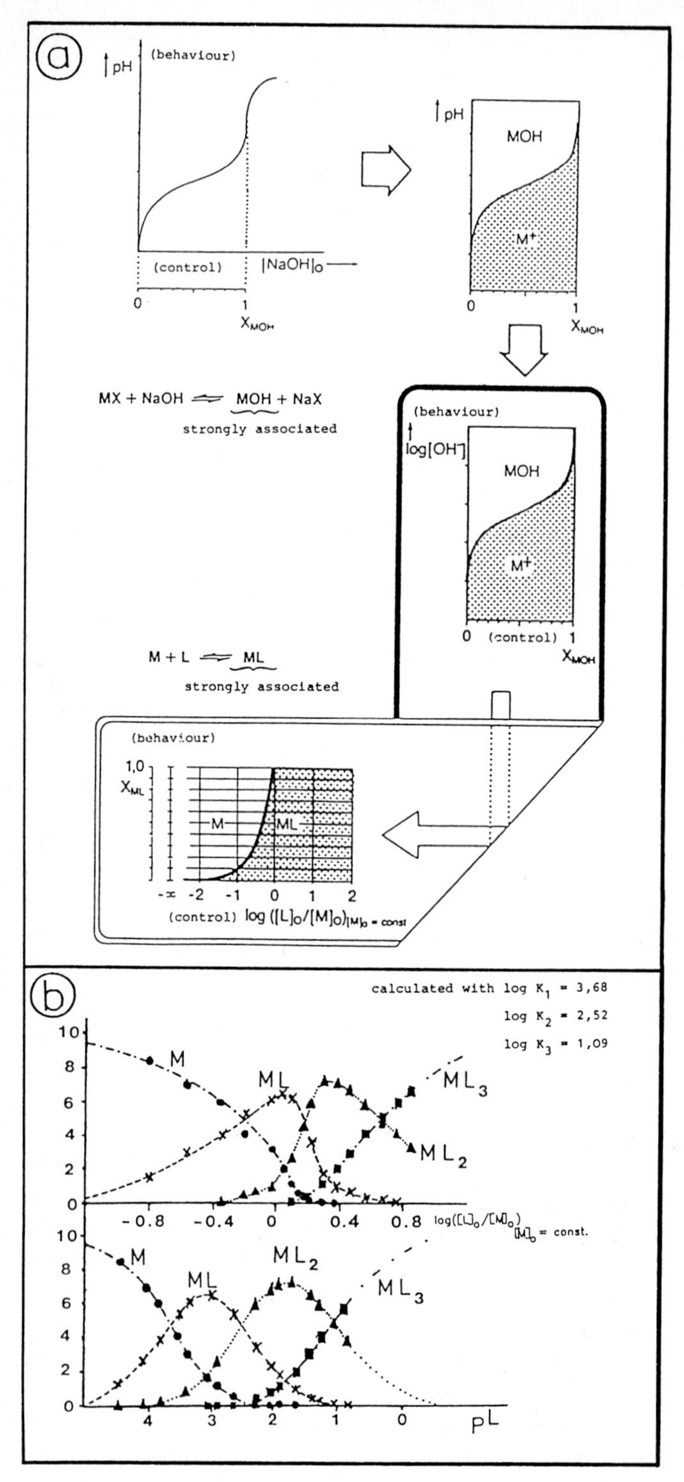

a
pH
(behaviour)
(control)
|NaOH|o
0
1
XMOH
pH
MOH
M+
0
1
XMOH
MX + NaOH ⇌ MOH + NaX
strongly associated
(behaviour)
log[OH-]
MOH
M+
0 (control) 1
XMOH
M + L ⇌ ML
strongly associated
(behaviour)
1,0
XML
M
ML
-∞ -2 -1 0 1 2
(control) log ([L]o/[M]o)|M|o = const
b
calculated with log K1 = 3,68
log K2 = 2,52
log K3 = 1,09
M
ML
ML3
ML2
-0.8 -0.4 0 0.4 0.8
log([L]o/[M]o) [M]o = const.
M
ML
ML2
ML3
4 3 2 1 0
pL

Fig. 6.2a,b.

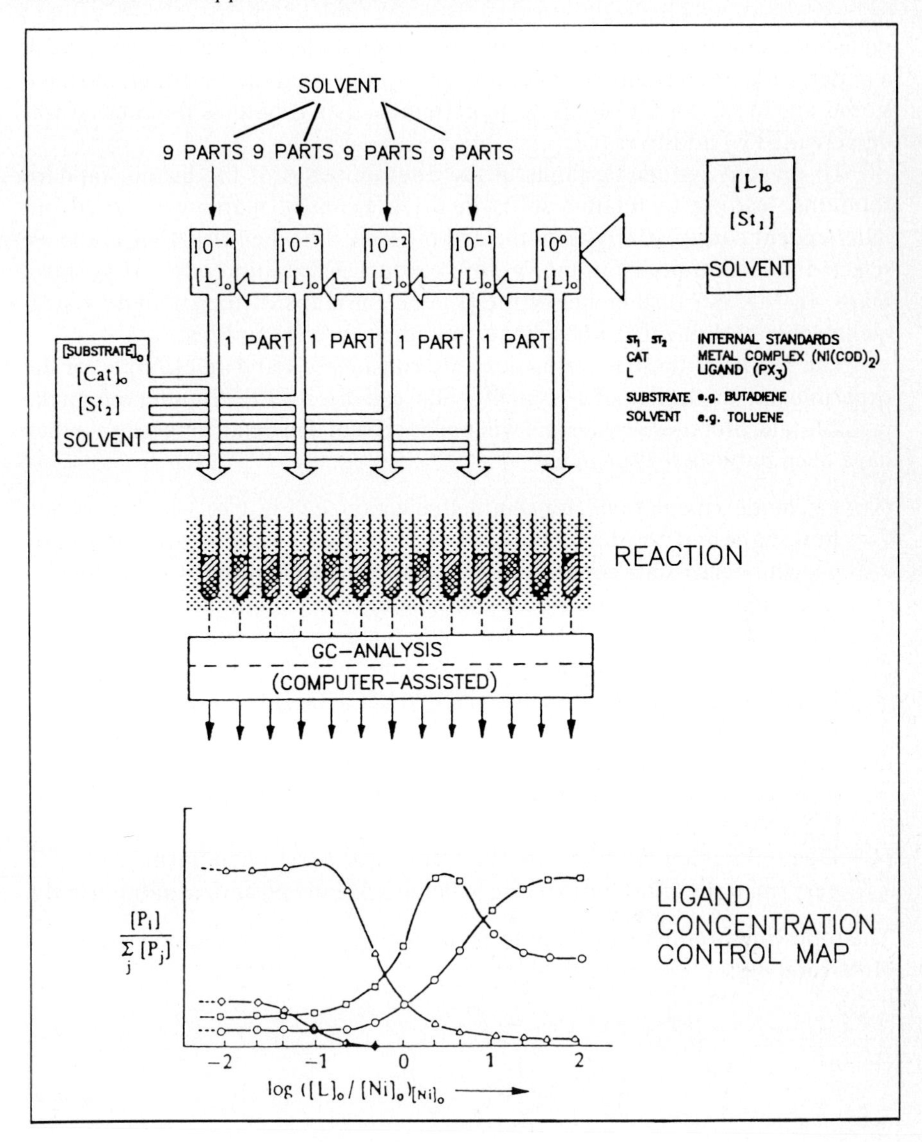

Fig. 6.3. Schematic representation of the procedure in the discontinuous INVERSE TITRATION of metal-catalyzed systems with ligand control

Fig. 6.2a. Comparison of continuous titration and discontinuous INVERSE TITRATION (vide text). **(b)** A comparison of patterns due to the thermodynamic coupling of three complexes with a different degree of association by plotting either as in the INVERSE TITRATION or as normally done sealed hermetically under an inert gas atmosphere, placed for a constant time in a water bath of constant temperature during the catalytic reaction, and then stored at −78 °C for GC analysis. In extremely active systems the catalyst was desactivated by additives.

To prevent systematic faults in the dilution series of the ligand standard solutions, leading to relative shifts in the [L]-control maps, we carried out independent control catalysis on the 250-ml-scale. For the ($[L]_0/[Ni]_0$) ratio we selected inflection points in the varying product distribution of the [L]-control maps. In Fig. 6.6 further below the [L]-control maps of the catalytic system nickel/phenyl-diphenoxi-phosphane/ butadiene are exemplified.

The results of these investigations are considered to be an example for the experimental evolution of a system (vide e.g. Fig. 6.12). Simulations make possible interpretations easier and will not be repeated in detail here because they have been published elsewhere:

(1): e.g. model calculations (for a limited selection see Fig. 6.4), which work out, how shape and position of the curves change by activation/inhibition, or by varying steady state concentrations in subsystems, in which – per definition (see Chap. 7) – the individual components are in equilibrium [19, 63, 70, 201].

(2): A "product stream diagram" (vide Scheme 3.2–3 in [39]) seems to be very helpful to visualize the structure of the THROUGHPUT, especially to avoid overinterpretation derived from partial concentration control maps.

(3): "Buffer curve diagrams" of different subsystems (vide Scheme 3.2–5 in [39]) give an idea of the range of existence for intermediates on the $[L]_0/[Ni]_0$-scale.

(4): Due to kinetic separation we can trace association phenomena even with very small $[L]_0/[Ni]_0$ ratios (vide Scheme 3.4–1 in [39] and in addition [63]).

The following products (2–7) are formed amongst others by oligomerization at $Ni^{(0)}$-catalysts:

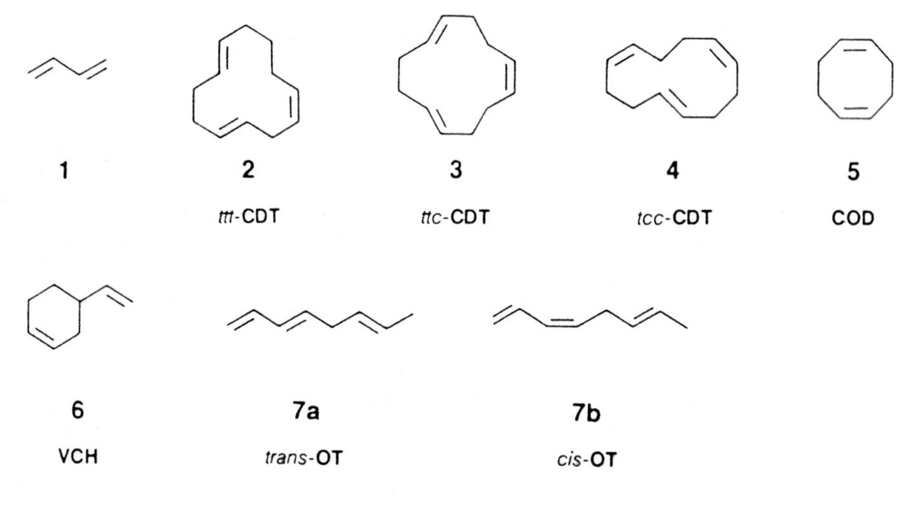

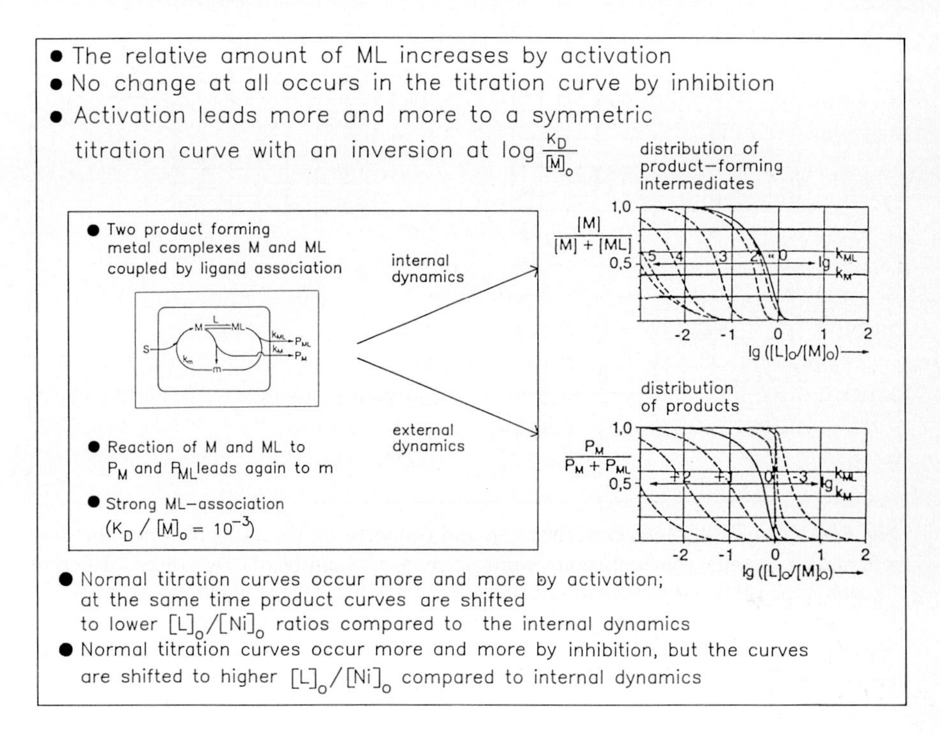

Fig. 6.4. A selected example of mathematical modelling

"The Art of Titration" [47] postulates that the used technique depends on the problem. Three possible types of [L]-control maps are published elsewhere [19,63] for the catalytic three-component system nickel/diphenylphenoxiphosphane/butadiene.

These three types of [L]-control maps give a striking reminder of the important rôle of Ostwald's dilution law in homogeneous catalysis [19].

N.B.: Impurities can become of great importance in such high excesses and effect changes in selectivities, which are caused only by these traces of unknown impurities.

In the discussion of the property-specific control of a directing ligand in homogeneous transition-metal catalysis, one has to be confident that the results of the considered experiments are comparable. Three different methods for analysis of the property dependent ligand control are schematically presented in Fig. 6.5 depicting three hypothetical concentration control maps of ligands with total different behavior in association.

Type a: Comparing the product distribution for different ligands at constant $([L]_0/[Ni]_0)[Ni]_0$ ratio denoted in all three maps by symbol (a), one has to take into account that the resulting selectivity can originate in different types of intermediates. For the given $[L]_0/[Ni]_0$ ratio, the first schematic [L]-control map

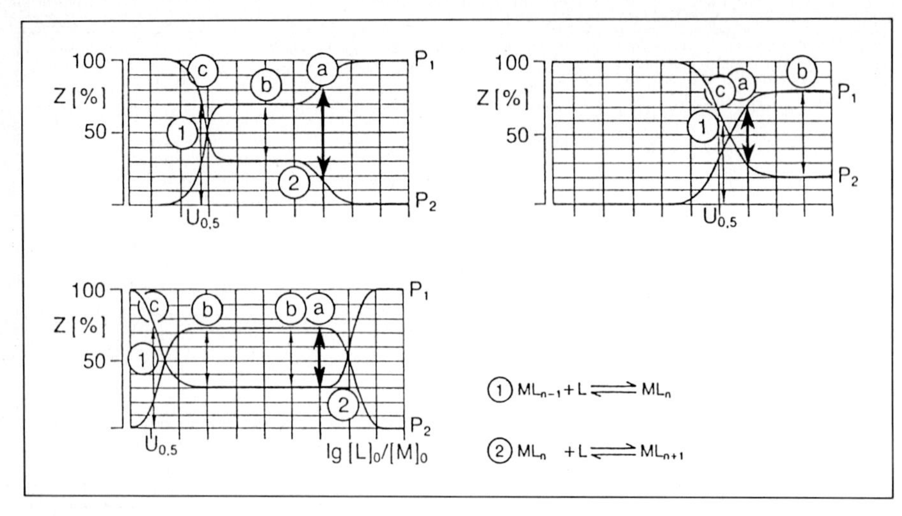

Fig. 6.5. Relation between concentration and property control of ligands presented by schematic ligand concentration control maps exemplified by three different ligands: type ⓐ Comparison of selectivity Z for a constant lg $[L]_0/[M]_0$ ratio. Type ⓑ Comparison of selectivity Z of metal complexes with identical degrees of association. Type ⓒ Comparison of the position of the titration curves on the scale caused by ligand properties e.g. in any case for the first association process (half tipover = $U_{0.5}$)

reflects the influence of the ligand in the second association step. In the middle one, this ratio is within the range of the first association step. In the last map, the resulting selectivity reflects the competition of the monoligand associates with one another. Without this information from [L]-control maps, an interpretation of the results for different ligands at constant $[L]_0/[M]_0$ ratio may be without any predictive power.

Type b: If a comparison of the chemical behavior for intermediates of the same degree of ligand association is desired, the accompanying [L]-control maps must be investigated in order to elucidate the range of existence of the intermediate complexes with comparable metal:ligand-ratio. Within this range, the kinetic selectivity is automatically described in the three maps denoted by (b).

Type c: In [L]-control maps the substitution of one ligand by another one might result in a change of the range of existence of the manifold intermediates. This change can be expressed by the ligand-property induced shift of the titration curves identified by the relative position of their inflection points $L_{0.5}$ (vide [63]) on the log ($[L]_0/[Ni]_0$) scale. These characteristic shifts provide information that is designated by (c).

This last discussion hopefully demonstrates the direct coupling of property- and concentration-control in unevolved complex chemical systems. Problems in the interpretation of [L]-control maps may also occur when the reaction rates of association and deassociation processes of ligands are slower than the observed metalcatalyzed organic syntheses (vide Fig. 6.15).

6.2 Examples for the Application of INVERSE TITRATION

We would like to demonstrate the broad application of the method by showing which different types of questions may be easier answered qualitatively and quantitatively by this method of INVERSE TITRATION in metal-catalyzed processes and in organic synthesis.

6.2.1 Application of INVERSE TITRATION in Metal Catalysis

First example in Fig. 6.6 and first question: What information can be inferred from different types of maps under coherent conditions by application of the discontinuous method of INVERSE TITRATION in the system $Ni^{(0)}$/butadiene/phenyl-diphenoxi-phosphane?

To verify the importance of so-called ligand concentration control maps we have chosen the above system, in which – on total conversion of butadiene – the products 2–7 are formed in varying proportions depending on relative and absolute ligand: metal ratio. Depending on this ratio one can recognize some association phenomena. The analysis by gaschromatography is carried out by choice of reaction time so that no educt remains, because when no very volatile educt is present this is easier. However, important information on activation/inhibition phenomena vanishes in such maps. It would be better to have product/educt maps at an uncompleted conversion stage.

In order to get more information, how type and number of differentiated reaction pathways may be influenced, we have extracted partial control maps from the original overall map i.e. (a) and (b). Thus in Fig. 6.6c to e the following maps are presented: (c): the extent of oligomerization, (d): distribution of dimers and (e): distribution of cyclotrimers.

In the case of the oligomer map, two very typical titration curves (B,C) are obtained. Their positions on the log ($[L]_o/[Ni]_o$) scale indicate that the corresponding intermediates occur in high steady-state concentrations. In the dimer distribution map four association processes (A,B,D,E) can be recognized. The first one (A) indicates an association process for an intermediate at low steady-state concentration. Association process (B) corresponds to the first ligand control in the oligomer map. The trimer distribution map is the simplest one. One association process is recognizable as a typical titration curve. This corresponds to the first association process (A) in the dimer map. A closer examination of the original [L]-control map indicates an alteration in the speed of formation of the trimer *4* and the dimer *5* as the original change in the product pattern (no COD/VCH-control!).

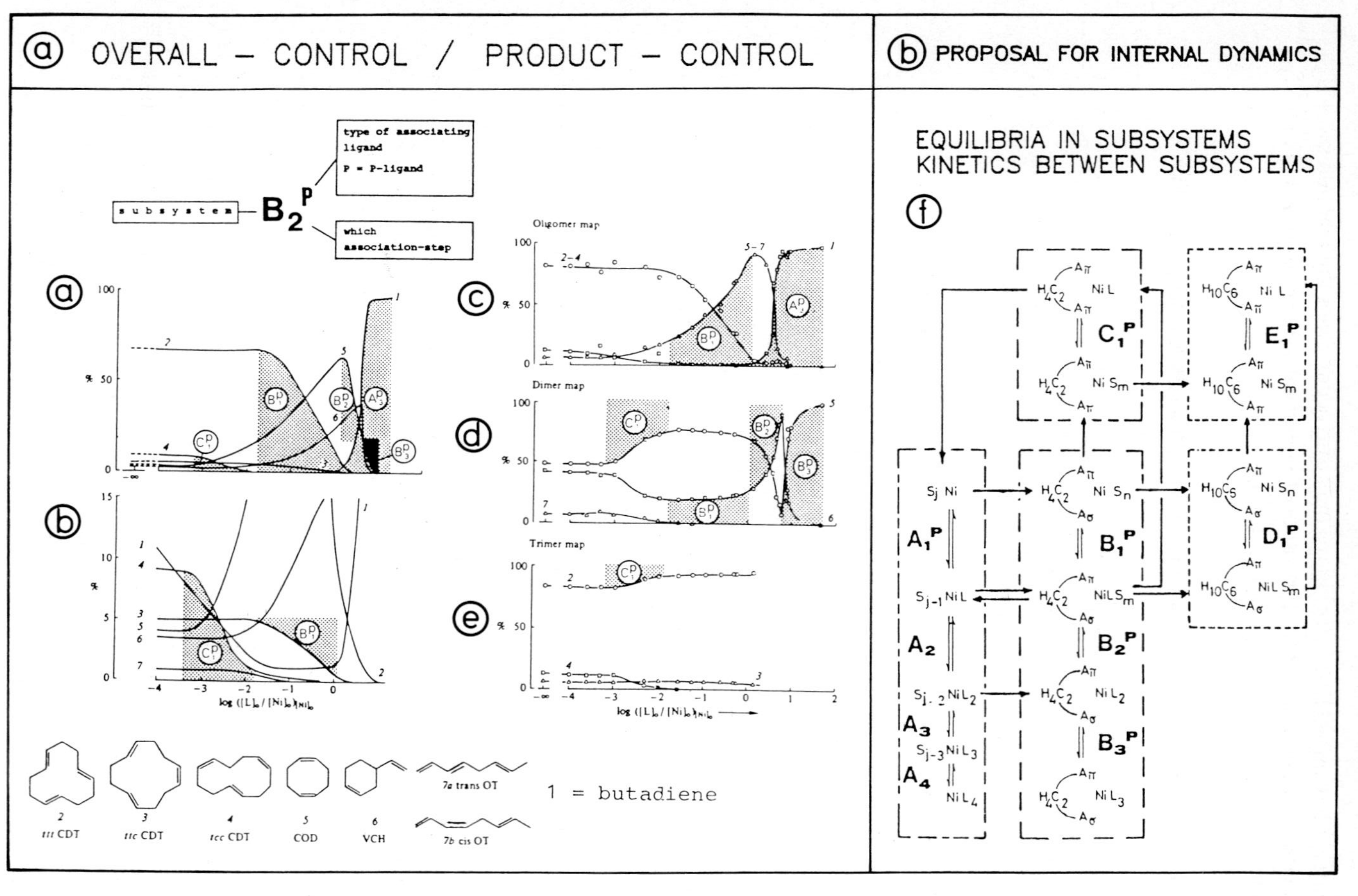

Fig. 6.6a-f. The products-educts maps **(a)**, a section out of this map **(b)** as well as selected information in form of oligomer **(c)**, dimer **(d)** and trimer distribution maps **(e)** are presented for the system $Ni(COD)_2/PhP(OPh)_2$/butadiene = 1:X:170(t_R = 15 h, 16 °C, $[Ni]_o$ 0,34 mmol/1 toluene). The structure for the THROUGHPUT is proposed in **(f)** (for further definitions vide Fig. 7.1)

The possible structure of the THROUGHPUT is added in Fig. 6.6f using information from "product stream" and "buffer curve" diagrams [39].

Second example and second question: How can we easily get information about thermodynamic competition between substrates?

In Fig. 6.7a the [L]-control maps for the four-components-system nickel/triphenylphosphane/butadiene/propene are given and this may be compared with the three-components-system nickel/triphenylphosphane/ butadiene [63,64]. The amount of the produced 2:1-co-oligomers of two butadienes and one propene is proportional to the cyclotrimers of the butadiene formed. The vanishing of the cyclotrimers and the 2:1-co-oligomers is directly coupled with the increase in cycloocta-1,5-diene. This suggests a subsystem, in which the cyclo-trimerization and dimerization of butadiene and the 2:1-co-oligomerization of butadiene and propene are commonly ruled by thermodynamically controlled competition of butadiene, propene and the directing ligand. The partial co-oligomer map of distribution of the individual 2:1-co-oligomers as a function of $\log([L]_0/[Ni]_0)$ in Fig. 6.7b indicates that two separated ligand association processes determine the ratio of CYCLIC to OPEN-chained products. Ligand association favours the formation of cyclic products. The existence of the two titration curves at different relative ligand-concentrations suggests at least two kinetically separated subsystems, the one forming the co-oligomers 8 and 9 and the other 10 and 11. The isomer 12 is formed independently.

Third example and third question: Can we have a similar control in catalytic processes, although starting from different metal-complexes?

For the dimerization of propene by homogeneous nickel catalysts e.g. two types of starting complexes are used, nickel-olefine-complexes like bis-cycloocta-1,5-diene-nickel [202] and allyl-halo-nickel complexes like bis-(π-allyl-chloro-nickel) [203,204]. The latter leads to a more active catalyst; using the former, an incubation period after mixing of the components has to be accepted. We were interested here in the question, whether the results of the catalytic systems are comparable or not. The dimers formed have three different skeletons: hexenes, methylpentenes and 2,3-dimethyl-butenes (but see Fig. 8.3). We investigated the catalytic systems $Ni(COD)_2/AlEtCl_2$/tricyclohexylphosphane/ propene (1:4:×:200) and allyl-NiCl/$AlEtCl_2$/tricyclohexylphosphane/propene (1:4:×:200) in liquid propene and chlorobenzene. The resulting [L]-control maps are shown in Fig. 6.8. We have limited the Ni:propene ratio to no more than 1:200 to be able to keep the reaction temperature constant as far as possible [39].

The dimers are listed under the three possible skeletons dimethylbutane 2-methylpentane and hexane ignoring the different isomerization properties of the two catalytic systems. As it can easily be seen there is no detectable difference in the skeleton-determining step between both systems from the standpoint of the [L]-control maps. Only one product-determining ligand-association process can be recognized: the range of existence of L-association and the changing product

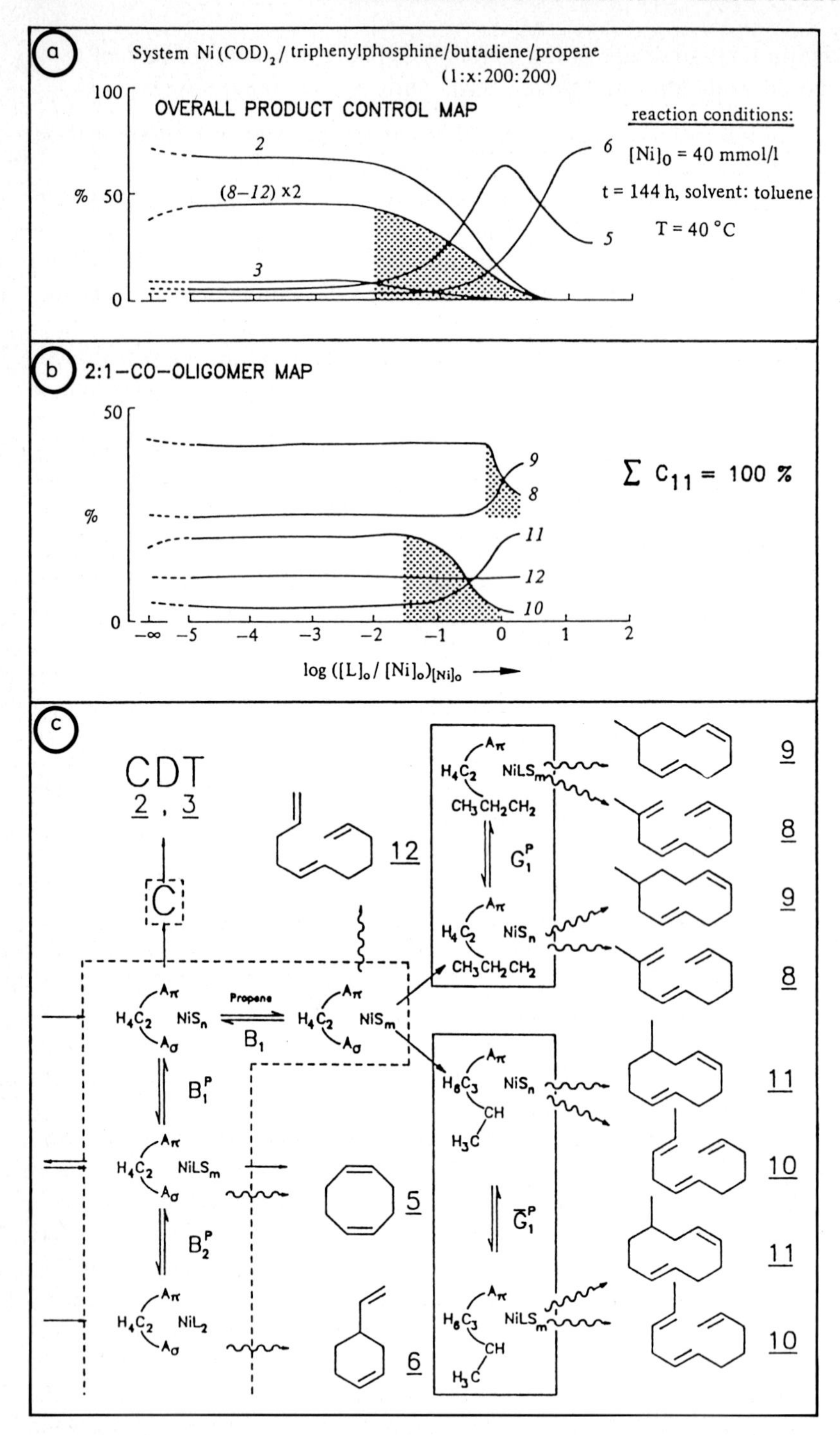

Fig. 6.7a-c. The influence of TPP concentration is presented for **(a)** the product map and **(b)** the 2:1-cooligomers for the THROUGHPUT is proposed in **(c)**

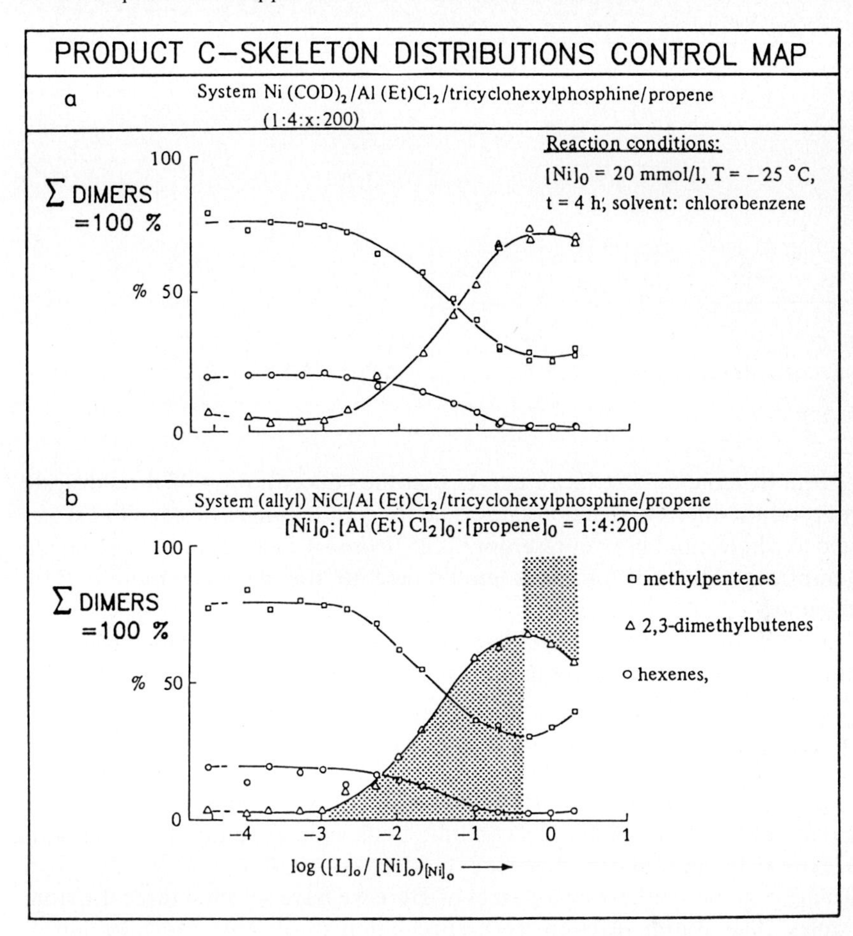

Fig. 6.8a,b. Here the similar control by the amount of the ligand is presented by comparison of two product (skeleton distribution) maps with different starting complexes

pattern being the same for both starting complexes. But we can not conclude from the map, that we have identical catalysts. One of them may be a Lewis-type system like $RX_2Al \leftarrow NiL_x$ and the other a Brönstedt-type system like $[H \leftarrow NiL_x]^+$. These two systems – or the same catalysts formed through different pathways – could be influenced by ligand in a similar manner. A first and a second step in the activation of propene, which are kinetically separated as in Fig. 6.7b, can not be identified [203,205]. A decision about the cooperativity displayed by the number of propene molecules in wide ranges of substrate concentration in this metal catalyzed system could only be made by applying an enzyme formalism determining the Hill coefficient.

Such investigations have already been done for the catalytic cyclodimerization of butadiene at $Ni^{(0)}$-tri-*o*-phenylphenyl-phosphite-(1:1)-catalysts (Hill coefficient = 2) and the cyclotrimerization at $Ni^{(0)}$-catalysts (Hill coefficient = 1) [63].

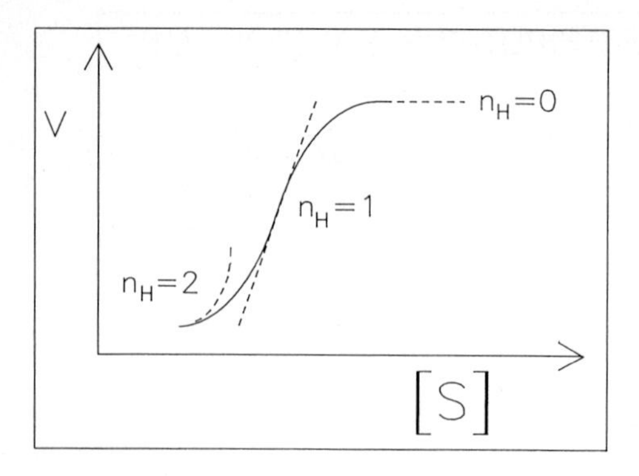

Scheme 6.2

From Scheme 6.2 the fact emerges that the concentration of the substrate might produce different reaction rates valid for certain ranges (broken lines) and related to whole numbers (representing 2., 1., 0. order). Representation of data in the transient ranges is of great importance to the determination of Hill coefficients.

In the Figures 5.7–10 we have already discussed an example of metal induced reactions and other investigations have been published elsewhere [61,62,188].

6.2.2 Application of the INVERSE TITRATION in Organic Synthesis

We developed the discontinuous method of INVERSE TITRATION for metal-catalyzed processes. However with time we have learned that this method is also extremely helpful for detecting hidden associations and couplings in "normal" organic syntheses very quickly. Here we have selected three different examples: The Wittig reaction, H-X-elimination to olefines and nitration of benzene-derivatives.

First example in Fig. 6.9 and question: Why do we observe the so-called "salt-effect" [206] that controls *cis-/trans*-selectivity of olefines in the Wittig reaction, although we know that the salt – formed in the course of ylide synthesis – associates to triphenylphosphaneoxide formed during the final process?

First of all let us consider the "normal" situation in Wittig reactions. The ylide may be synthesized by reacting e.g. a phosphonium-iodide with a base, in this case a LiOMe. Then among the coloured ylide, one equivalent of LiI and one equivalent of MeOH is present in the solvent, too. Therefore we purified the ylide by recrystallization and – as shown in Fig. 6.9a and b – "titrated" discontinuously the two-components-system ylide/aldehyde (1:1) with LiI and in the three-components-system ylide/aldehyde/LiI (1:1:1) with triphenylphosphaneoxide [31]. This enabled us to unravel the origin of the salt effect. The increasing amount of e.g. LiI leads to full – catalytic – salt effect in understoichiometric amount (at $\log[\mathrm{LiI}]_0/[\mathrm{Ylide}]_0 = -0{,}5$). On the other hand the beginning association of added

LiI to triphenylphosphanoxide, that is formed in addition up to the stoichiometric amount during the process, can only be observed starting by a twofold excess of the added oxide. Both facts together – understoichiometric effect of LiI and overstoichiometric amount of triphenyl-phosphanoxide to bind LiI – make clear, that a full salt effect remains observable in this case in a one pot reaction.

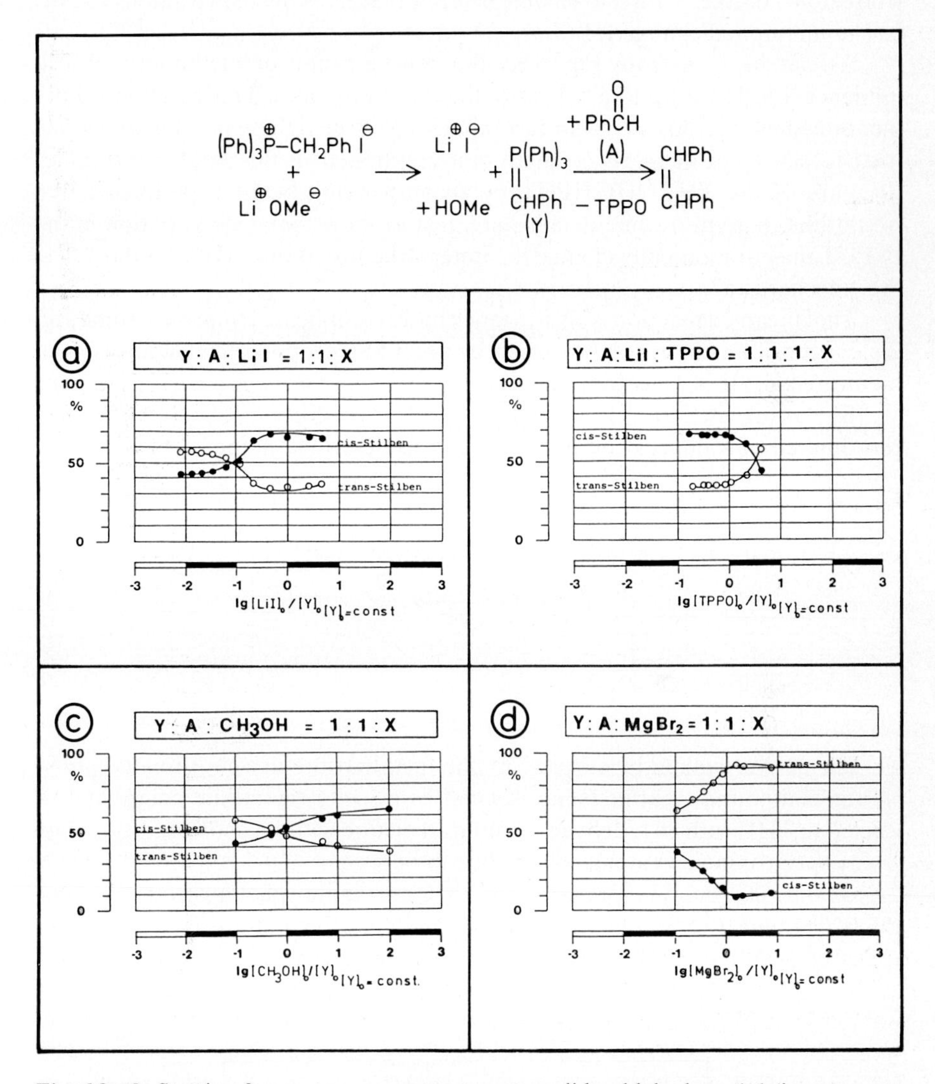

Fig. 6.9a-d. Starting from a two components system ylide:aldehyde = 1:1 (in every case 1 mmol in 3 ml THF solution at T = O °C and t_R = 1 h) the effect exerted by the amount of a third component is demonstrated (LiI **(a)**, methanol **(c)** and $MgBr_2$ **(d)**) or starting from a three components system ylide:aldehyde:LiI = 1:1:1 **(b)** the effect of a fourth component TPPO on the stereoselectivity of the olefine synthesis in the Wittig reaction

However, by adding a sixfold excess of TPPO salt-containing ylide solution is "salt-free" in this case by association of the salt with the TPPO!

When we "titrate" with $MgBr_2$ instead of LiI an alternative type of *cis-/trans*-selectivity in olefine formation is observable (Fig. 6.9d). This is in agreement with the PSE-sector rule 3 (change in anion I/Br) discussed earlier in Chap. 3 along with other examples. Nevertheless we must be careful. It could be a threefold change: 1. First to second group: Li/Mg: 2. One to two anions $I^-/2Br^-$ 3. The mentioned change I/Br.

As can be seen from Fig. 6.9c, the concentration of methanol will also influence a little bit up to a 1:1 ratio, the *cis-/trans*-ratio. This titration is only mentioned in order to discuss an accompanying effect of these investigations. We tried to clarify in detail – normally not mentioned in this book – the inner structure of the THROUGHPUT by accompanying NMR-experiments. We would like to go into some details here, just to show that each variation in the INPUT may dramatically change the inner structure of the THROUGHPUT – the "mechanism".

That means, a reaction is always much more complicated than we assume and the danger of misinterpretation exists by accepting proposed mechanisms from literature [33,34]. *A "mechanism" is always only true for the investigated system!*

By adding methanol to the isolated ylide, an addition to a colourless five coordinated phosphorus compound occurs, as described in the literature [207].

$$Ph_3P{=}CH_2 + CH_3OH \rightleftharpoons (CH_3O)(CH_3)PPh_3$$

But we were taken by surprise by the fact, that the five coordinated phosphorus compound – after being isolated as a white crystalline compound – reacted in THF with an aldehyde forming an olefine without showing any colour or any spectroscopic evidence of an ylide in between.

On the other hand we prepared, at −78 °C an oxaphosphetane from ylide and aldehyde in a 1:1-ratio and added one equivalent of methanol. Two new peaks appeared in the ^{31}P-NMR-spectrum (vide Scheme 6.3). An unsolved problem still

apical

$$\text{Ph}_3\text{P}(\text{O}{-}\text{CHR}{-}\text{CH}_2) \rightleftharpoons \text{Ph}_3\text{P}^{+}(\text{O}^{-})\text{CH}_2{-}\text{CHR}^{\oplus} \rightleftharpoons \text{Ph}_3\text{P}(\text{CH}_2{-}\text{CHR}{-}\text{O})$$

equatorial

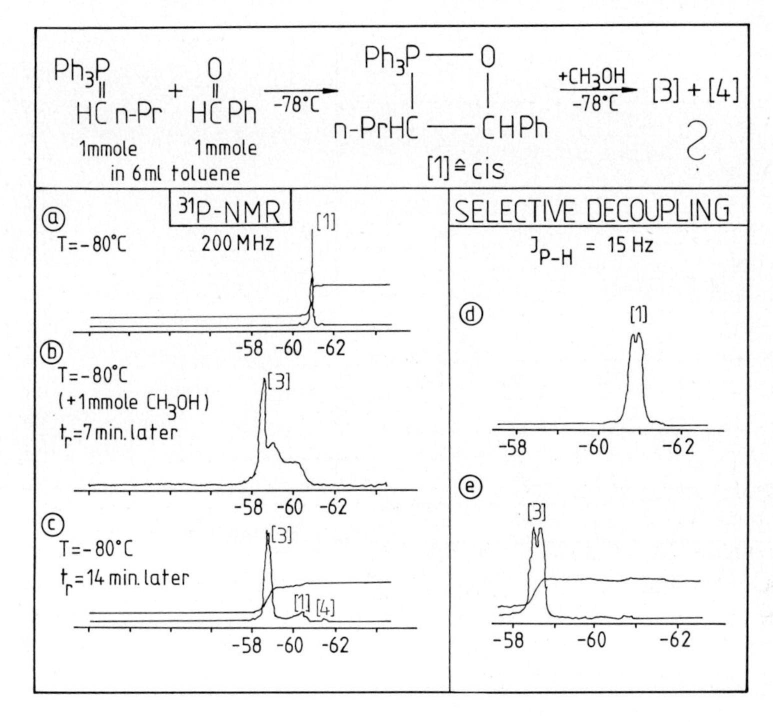

Scheme 6.3

remains whether an association or an addition of methanol or a rearrangement within the oxaphosphetanes take place. A possible and theoretically necessary rearrangement is published without actually experimental realization [33].

By adding – on the other hand – one equivalent of LiI to oxaphosphetanes, coalescence phenomena will occur effected by betain-formation perhaps via LiI-addition (vide Scheme 6.4).

Without discussing further details, we hope to convince the reader that "mechanisms" (or better "populations of intermediates") may dramatically change by variations in the INPUT and that the discontinuous method of INVERSE TITRATION immediately identifies hidden associations. The same remark is true for the next example.

Second example in Fig. 6.10 and questions: Is the variation in the alkyl-groups of a base (increasing "steric" – better repulsive – effect) really responsible for Δ1-/Δ2-control in olefine formation from *tertiary* alkylhalides? Has the solvent any influence?

Δ1-/Δ2-olefine synthesis is controlled amongst others by the type of alkyl-groups in alkoxides according to [177] (vide Scheme 5.2).

Usually textbooks do not account for the fact that in the real experiments [176] the reaction is always carried out in the corresponding alcohols. The

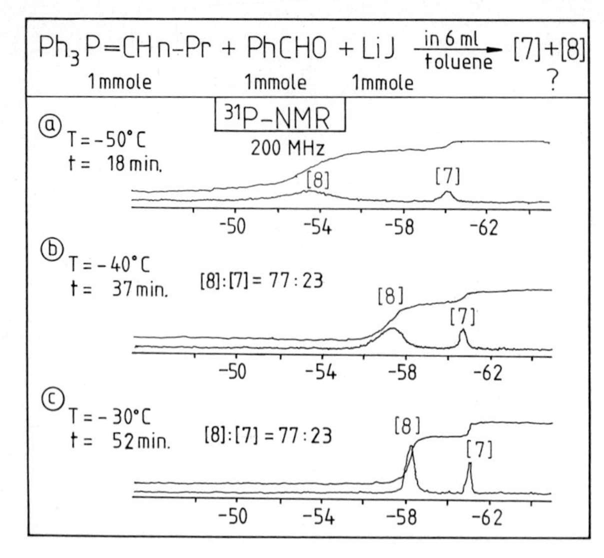

Scheme 6.4

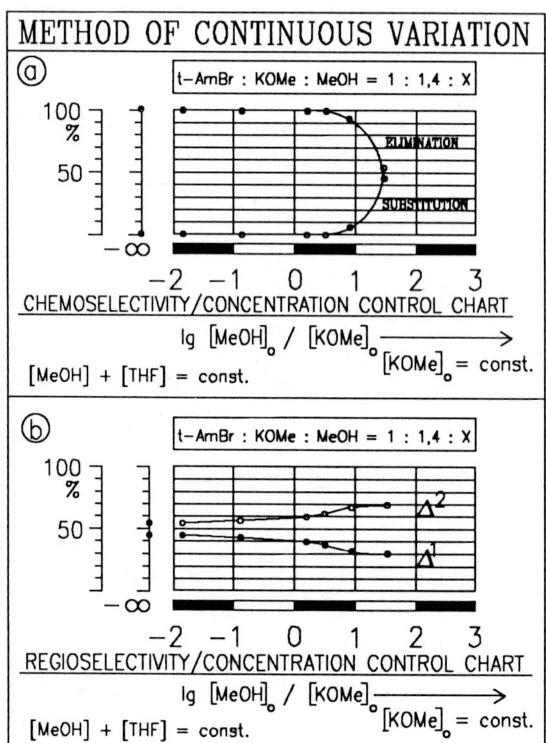

Fig. 6.10a,b. By applying the method of continuous variation the sum of the concentrations of THF and the alcohols remains constant in addition. **(a)** demonstrates the chemoselectivity and **(b)** the stereoselectivity of olefine formation by the increasing exchange of THF as solvent by methanol

influence displayed by the series Me/Et/*i*-Pr/*t*-Bu depends alternatively on the system (Na/K as counterions in the iodide systems: vide Fig. 5.2).

The overwhelming influence of the solvents R^x-OH (in addition $\approx$ 30% difference) in Δ1-/Δ2-control was easily verified by the discontinuous method of continuous variation [179] as shown in Fig. 6.10. In the three components system THF was replaced in discrete steps by methanol. Looking at the product map (a) we realized that increasing concentration of the alcohol even favours substitution reaction using primary alcohols as solvents. At the same time we could clearly identify in the partial olefine selectivity map (b) the overwhelming influence of the solvent on Δ1-/Δ2-olefine formation. That demonstrates the solvent control (control effect in addition $\approx$ 30%) to be more important than the alkyl influence of the base (control effect $\approx$ 10%).

Third example and question: Is there any experimental evidence for separated *ortho-/meta-* and *para-/meta*-coupling in the electrophilic substitution of mono-substituted benzene-derivatives?

Clearcut evidence for separated couplings of the discussed types is shown in Fig. 6.11. As described in the literature [208] the addition of Hg^{++}-salts influences the positional isomerism in the nitration of monosubstituted benzene-derivatives. One disadvantage is the low chemoselectivity of mono-nitrated compounds in this system. But especially in the case of *t*-Bu-substituted benzene we observed higher chemoselectivity (max. 45%) [35].

Therefore we "titrated" the system *t*-Bu substituted benzene/HNO_3/glacial acetic acid with $Hg(NO_3)_2$ [35].

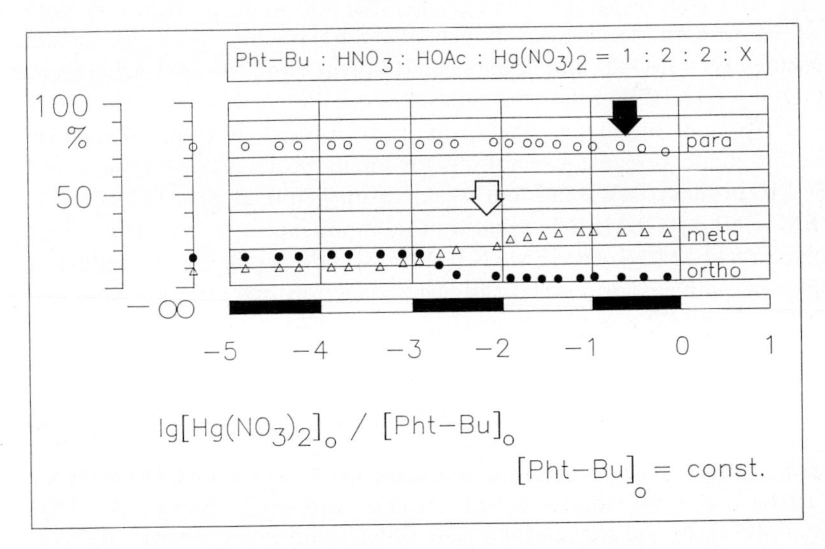

Fig. 6.11. Influence of the amount of Hg^{II}-nitrate on the *ortho-/meta-/para*-selectivity during nitration of *t*-Bu-benzene (at 50 °C and t_R = 24 h) under coherent conditions (65% HNO_3)

6.2.3 Application of INVERSE TITRATION in SYSTEM ENLARGEMENT

Question to Fig. 6.12: Can the relative amount of two ligands cooperate in such a way that two different products are formed in high selectivity? Yes.

Here the alternatives MUCH/A LITTLE as amount and DO/ACC at the central atom display a cooperating or compensating effect! In Fig. 6.12 an example is given for the formation of two different main products – CYCLIC and OPEN dimers of butadiene – as a consequence of alternative ways leading from the two components system to the four components system in the same solvent. The *sequence of intermolecular SYSTEM ENLARGEMENT* (see Fig. 6.12a and (b) or (c) and (d)) is *very important* for deciding which type of reaction will be optimized. The same result of cooperation of two ligand concentrations is shown once again by the hyperplanes in Fig. 6.12e. A preliminary rationale of the THROUGHPUT by these observations is given in [13]. We wish to point out two important problems of this rationale:

(1): The influence of DO/ACC heteroatoms (N/P) representing central atoms of the ligands on the ALTERNATIVES in the products (CYCLIC/OPEN) and
(2): the cooperativity of alternatives of amount – reduced to MUCH/A LITTLE – and of properties via DO/ACC heteroatoms as central atoms of ligands we have already referred to in Fig. 4.6d.

Critically one could object, that in morpholine and triphenylphosphite totally different substituents are bound to the central atoms N/P. This objection seems to be justified, when we consider the results of Fig. 6.13 in the product distribution of the three components system $Ni^{(0)}$/ligand/butadiene with the alternative ligands HN(Me)Ph/HP(Me)Ph. The ligands differ only in the central atoms with identical substituents. The main products of the butadiene dimers still remain *n*-octatrienes (N = n-OT) and cycloocta-1,5-diene and 4-vinylcyclohexene (P = COD + VCH). But at the "N"-catalyst about 45% of C_8-alkylated amines (2:1-adducts out of butadienes and amine) are formed. But an unequivocal statement is made possible concerning the main products (N = OPEN and P = CYCLIC product) only by adding in each case 0,1 equivalent of TPP per $Ni^{(0)}$ (SYSTEM ENLARGEMENT to a four components system). This intermolecular SYSTEM ENLARGEMENT suppresses the N-alkylation product by activating the pathways to *n*-OT [209]. The concerted processes are favoured

Fig. 6.12a-f. Different procedures for the intermolecular SYSTEM ENLARGEMENT from two (basic system: $Ni^{(0)}$/butadiene) to four components systems. Stepwise SYSTEM ENLARGEMENT by: **(a)** triphenylphosphite (three components systems); **(b)** morpholine (four components system), **(c)** morpholine (three components system); **(d)** triphenylphosphite (four components system). Twofold SYSTEM ENLARGEMENT by: **(e)** morpholine and **(f)** triphenylphosphite ($t_R = 6$ h; $T = 60\,°C$)

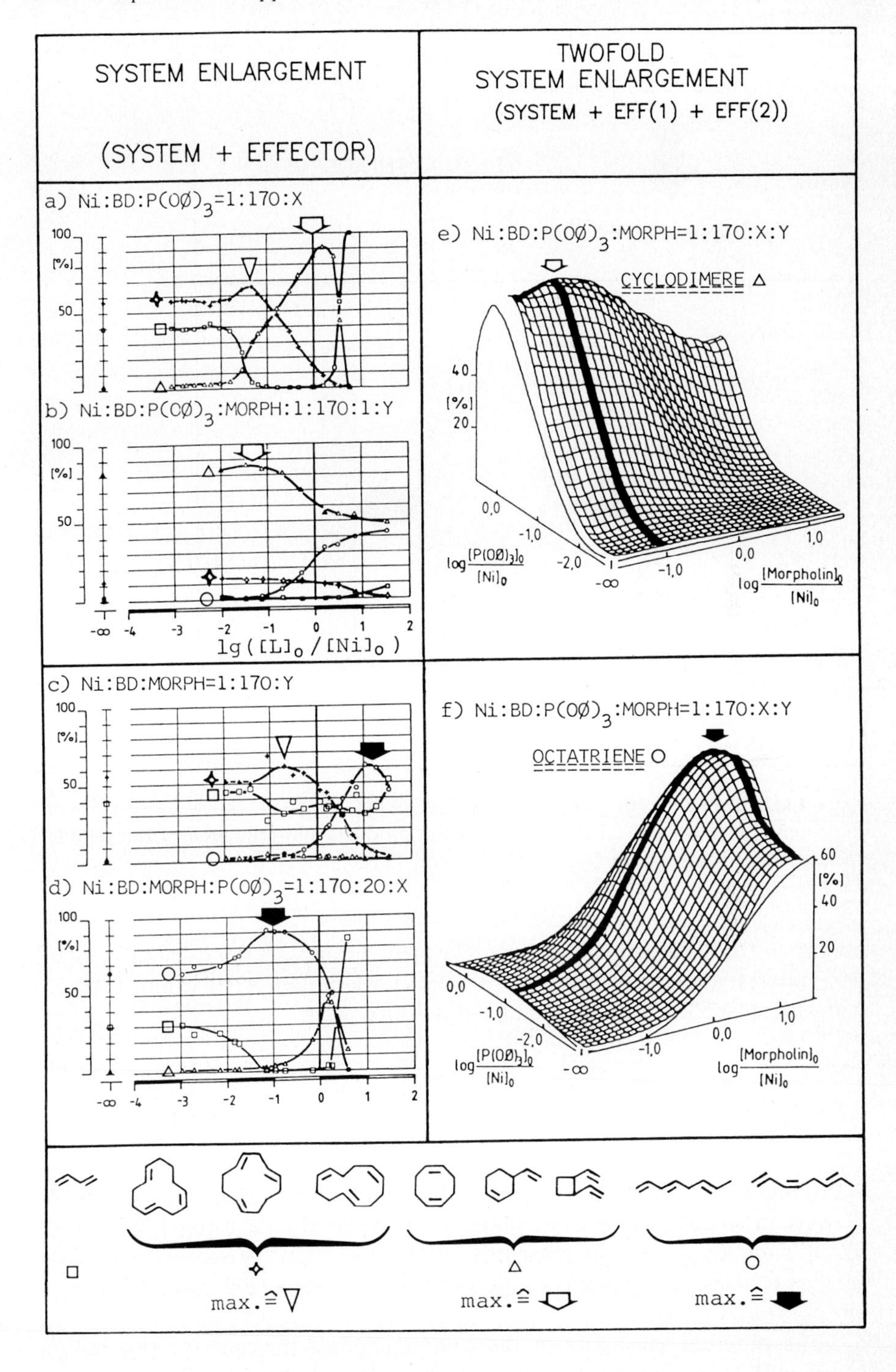
SYSTEM ENLARGEMENT
(SYSTEM + EFFECTOR)
TWOFOLD SYSTEM ENLARGEMENT
(SYSTEM + EFF(1) + EFF(2))
a) Ni:BD:P(OØ)3=1:170:X
b) Ni:BD:P(OØ)3:MORPH:1:170:1:Y
lg([L]o/[Ni]o)
c) Ni:BD:MORPH=1:170:Y
d) Ni:BD:MORPH:P(OØ)3=1:170:20:X
e) Ni:BD:P(OØ)3:MORPH=1:170:X:Y
CYCLODIMERE
log [P(OØ)3]o/[Ni]o
log [Morpholin]o/[Ni]o
f) Ni:BD:P(OØ)3:MORPH=1:170:X:Y
OCTATRIENE
max.≙
max.≙
max.≙

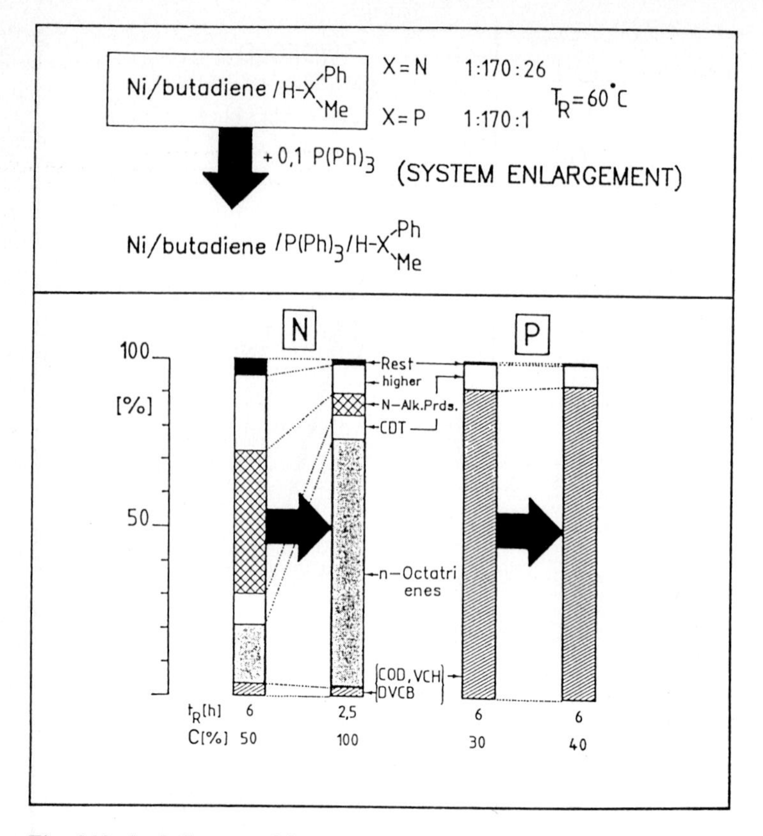

Fig. 6.13a,b. Influence of the central atoms N/P of ligands with identical substituents on the chemoselectivity (CYCLIC/OPEN) of the products and reduction of the product spectrum by adding 0,1 equivalent of the ligand TPP

perhaps by the greater polarisibility of the phenylgroups. In this discussion it is of great interest, that the so-called 1,3-dipolar cycloadditions [210] can be realized with special success with phenylsubstituted substrates.

6.2.4 Application of INVERSE TITRATION for the Proof of the Influence of Catalyst Poisons

According to a widespread prejudice sulfur compounds are generally said to be poisons for Ni-catalysts, because black NiS is expected to be formed. We would like to show by the results presented in Fig. 6.14, that the sensitivity against catalyst poisons as inhibitors is dependent on the system [209] as in the case of effectors (activators).

The different sensitivity of the catalyst against the amount of a sulfur compound is demonstrated in (a) in both the maps of the three components

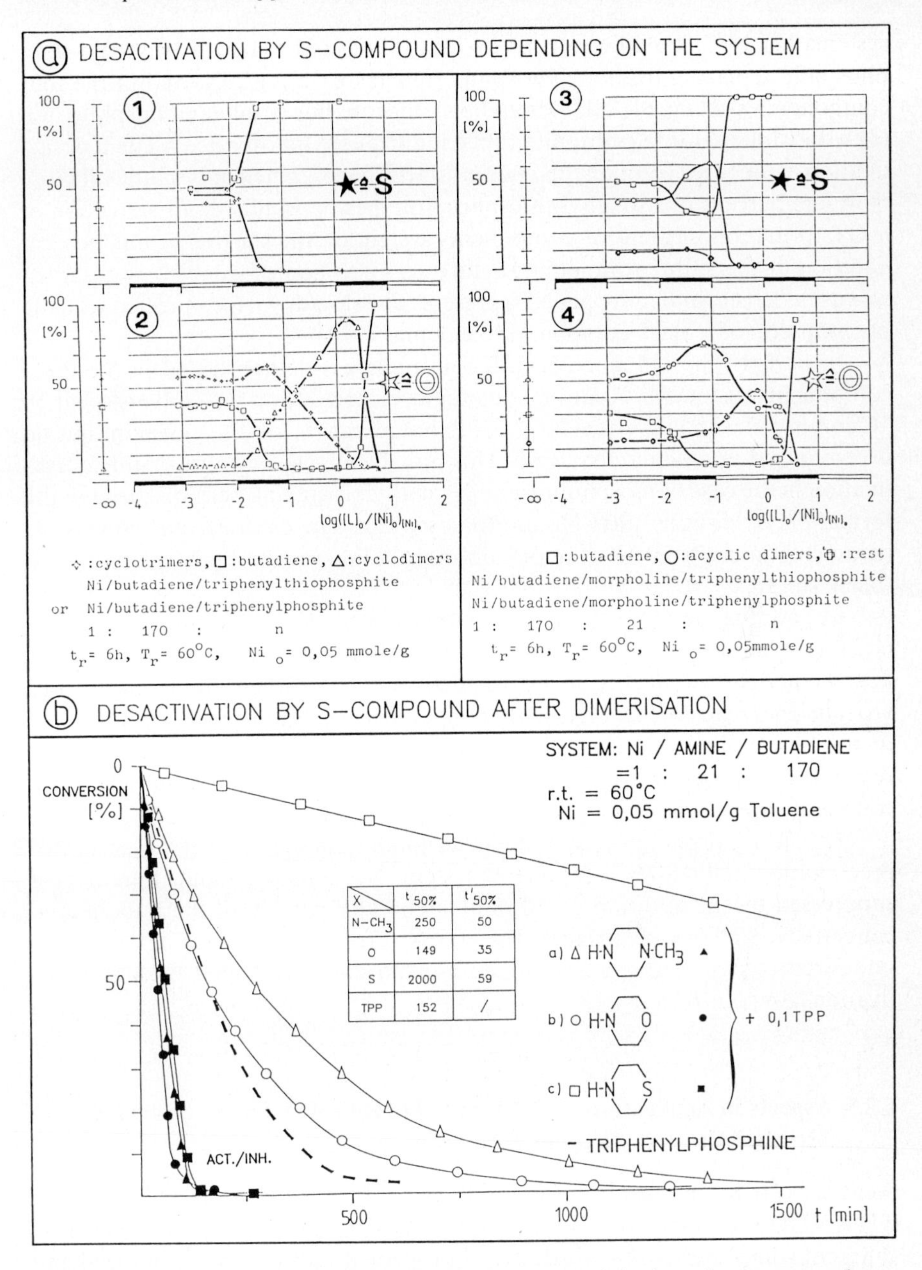

X	$t_{50\%}$	$t'_{50\%}$
N-CH3	250	50
O	149	35
S	2000	59
TPP	152	/

Fig. 6.14a,b. The desactivation by catalyst poisons depends both on the amount and type of the poisoning compound and on the composition of the system (vide context, too)

systems (2) $Ni^{(0)}$/butadiene/triphenylphosphite and (1) with triphenyl-thio-phosphite replacing triphenylphosphite (in each case 1:170:X) and in the four components systems (4) $Ni^{(0)}$/butadiene/morpholine/triphenylphosphite and (3) with triphenyl-thio-phosphite replacing triphenylphosphite. While a desactivation in the three components systems starts sharply when only a little bit more than one percent of triphenyl-thio-phosphite per Ni is added, an activation of *n*-OT-formation in the four components system occurs starting at one percent addition of the sulfur compound, before an inhibition of catalysis is observed when ten percent and more per Ni is present. The slow down is connected with the precipitation of a black compound (elementary Ni!).

In part (b) the influence of different amines is demonstrated in the three components system $Ni^{(0)}$/amine/butadiene (1:21:170) on the conversion at 60 °C compared with the system $Ni^{(0)}$/TPP/butadiene (1:1:170). A clear inhibition is recognized in all amine systems. This may be effected by a reversible desactivation in the case of morpholine and N-methylpiperazine perhaps through the formation of dimeric allyl-Ni-amide complexes. In contrast, an irreversible desactivation starts in the case of thio-morpholine caused by Ni precipitation (black solid).

By adding to the three components system $Ni^{(0)}$/amine/butadiene (1:21:170) only 0,1 TPP the reaction rates in all three systems increase and become nearly equal up to a conversion of around 97% of the butadiene. But the so formed *n*-octatrienes react further on to higher oligomers in a slower consecutive reaction in the morpholine and N-methylpiperazine systems. Therefore the catalytic reaction has to be interrupted after the fast volume contraction to avoid this consecutive reaction.

The thio-morpholine system behaves totally differently. The observed Ni-precipitation in the three components systems Ni/thiomorpholine/butadiene is suppressed by the addition of 0,1 equivalent TPP per Ni till the butadiene is converted to 97% of *n*-octatrienes. Then the Ni-precipitation occurs again and the undesired consecutive reaction of the *n*-octatrienes is avoided by autodesactivation [209].

6.2.5 Aspects in Application of INVERSE TITRATION for the Ni-ligand Modified Catalytic Propene Dimerization

As described by the results in Fig. 6.8 we applied the method of INVERSE TITRATION in order to demonstrate the similar ligand control starting from different complexes in the Ni-catalyzed propene dimerization. Ni-induced and -catalyzed processes are described extensively in [211].

We restricted our Ni:propene ratios of 1:200, in the experiments described above, to be able to control the chosen reaction temperature easily. In two further ligand maps we were surprised by the fact (vide Fig. 6.15a and b), that the first association phenomenon by adding tricyclohexylphosphane is smaller by three powers of ten. How can this discrepancy be explained? It is caused by the

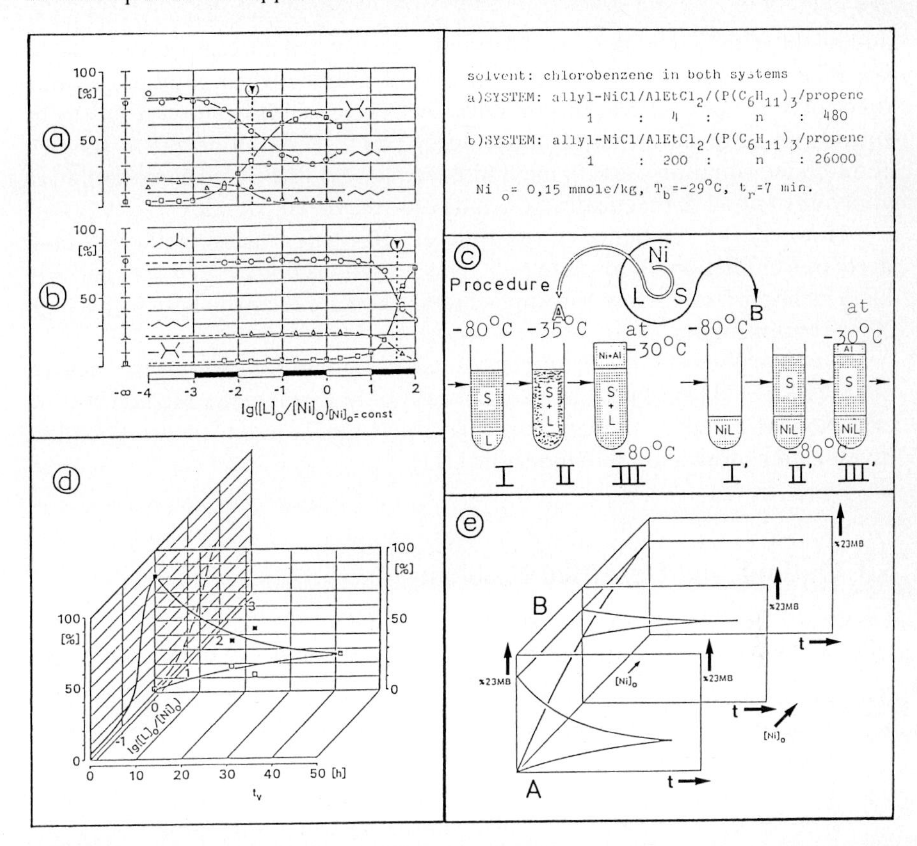

Fig. 6.15a-e. The investigations of influences by ligand concentrations on skeleton control show dramatic differences in **(a)** and **(b)** in different systems, which are determined by the experimental procedures depicted in **(c)**. How to account for the extent of differences e.g. by the choice of prereaction time **(d)** or total concentration of the catalyst **(e)** is schematically represented (vide text, too)

methodic procedure. We prepare a ligand/substrate solution and as an upper layer we add at low temperatures a Ni/Al standard solution (procedure A in Fig. 6.15c). Mixing at the desired reaction temperature immediately induces the rapid catalytic reaction (vide [205]). But by choosing procedure B (in Fig. 6.15c), i.e. when we prepare first of all a (Ni + L) standard solution and add, as an upper layer, propene and further on in the upmost layer, the standard solution of the aluminium compound as coeffector, we get – despite the conditions of system (*b*) – a ligand map, that resembles the map in Fig. 6.15a.

Let us try to explain by using the two schemes in Fig. 6.15d and e. The sum of the 2,3-dimethyl-butenes (small black boxes for method B and white ones for method A) as part of all dimers is obtained as shown in Fig. 6.15d, when one prepares four individual reaction mixtures of the system (*b*) (L:Ni = 1:1) without adding the $AlEtCl_2$ component, shakes vigorously when the temperature –45 °C is reached and adds the aluminium compound immediately or after prereaction times of 17,21 and 41 hours. Probably the degree of dissociation α has to be considered as being of major importance. The slow development of the real equilibrated situation starts in method B from an 1:1 associated complex and in applying method A practically from a ligand-free metal complex.

However, the influence of the time of prereaction using method A or B decreases in relation to absolute $Ni^{(0)}$-concentrations (vide Fig. 6.15e) and will loose its importance at low Ni:propene ratios, that are identical here with a high $Ni^{(0)}$-concentration.

We have discussed this example in detail to express quite clearly the method of INVERSE TITRATION as being always only a pragmatic method for the evolution of complex systems; but this allows quick, exact, physicochemical investigations which are indispensable [211].

6.3 Outlook and Unsettled Problems

Of course other controls may be investigated by the influence of amounts, e.g.:

(1) The associations of electrons (for this aspect vide [212]), e.g. different oxidation stages of the same transition metal: "titration" of a $Ni^{(0)}$ standard with Ni^{II}-complexes and so on.
(2) The associations of two solvents producing various orders (e.g. THF and diethylether: vide Fig. 3.16, which control the order of the products).

In both cases the so-called "method of continuous variation" [179] (the sum of both variable concentrations is constant)

$$[A]_0 + [B]_0 = \text{const}$$

may best prove these statements.

The application of dynamic and continuous titration techniques, [47] is desirable in metalorganic catalysis, too. But this is a task for physicochemists and the chemist involved in preparation is normally overtaxed. "Relaxation titration techniques, on the other hand, need highly sophisticated methods of dynamic measurement not possible with standard laboratory equipment" [47].

We have choosen in most of the presented examples of catalysis a longer reaction time. Therefore the low boiling educts butadiene and propene were usually totally converted in most of the step-by-step experiments and thereby did not lead to any problems in the routine GC analysis.

But we have to keep in mind, that a lot of information is lost, e.g.

(1) Very informative activation/inhibition phenomena disappear (as in the partial control maps!).

(2) Kinetically controlled intermediate products can not be identified, such as *cis*-1,2-divinylcyclobutane in the Ni-ligand catalyzed dimerization of butadiene, which can only be recognized [150] with conversions of the diene of less than 80%.

These complications are determined by the properties of subsystems and not by the method itself; they can be avoided e.g. by cooled GC inlet systems. Nevertheless the discontinuous method of INVERSE TITRATION can not replace careful kinetic investigations, but it can contribute to a more rapid finding of optimal conditions for such physicochemical measurements.

Investigating the influence of bases, acids, cosubstrates, solvents, catalysts, ligands and so on by some powers of ten it is absolutely *necessary to purify the components of the systems carefully*. It may happen, that the controlling effect is only caused by the unknown impurity!

7 Molecular Architecture: Some Definitions

Dann hat er die Teile in seiner Hand,
Fehlt, leider! nur das geistige Band.

Faust:Studierzimmer
Johann Wolfgang von Goethe

7.1 Structure of the Whole System

In the structural analysis of a highly complex system, initiation of thinking in terms of system theory [213,214] exerts organizing power. The whole system is purposely divided into INPUT, OUTPUT and the – at most – open system of the THROUGHPUT, often called "black box" (see Fig. 7.1). INPUT means the reservoirs of the quantitatively (concentrations) and qualitatively (properties) differentiated educts, solvents and catalysts (specificities), their energies and information as well as the given experimental conditions such as temperature. OUTPUT is the quantitative and qualitative consistency of the reaction mixture (selectivities) as well as its energy contents.

The type of couplings demarcates the elements or subsystems in the THROUGHPUT. For our purposes in chemistry the inner structure of the "black box" will be elucidated in the best way by defining the individual chemical compounds as elements.

Those elements of the THROUGHPUT, which are related to each other by thermodynamic equilibria, represent a part of the system or a subsystem. The individual subsystems are coupled together by kinetic processes.

The INPUT is related – depending on system – to the OUTPUT with varying connectivity of strength. The unsolved problem is, what kind and amount of information from the INPUT – what type of its ALTERNATIVE PRINCIPLES (see Chap. 8) – determines the order in the product spectrum of the OUTPUT (see Chaps. 3 to 5).

There are two ways of investigating the complete system. To find out direct correlations of the variations either introduced in the INPUT or of the obtained changes in the OUTPUT to the behavior or properties of elements or subsystems in the THROUGHPUT, one will have to poke about inside the "black box". Now we have to study the so-called internal dynamics directly, e.g. by spectroscopic and kinetic investigations. An extremely careful kinetic investigation of an optically active hydrogenation catalyst was done by J. Halpern and coworkers [215] and a different system was elucidated by highly sophisticated NMR-studies by J.M. Brown [216,217]. The overall processes are known to the last details; the gist of the problem in all such cases has to be seen in the fact, that the information gained in detail is connected with an extremely high investment of time and –

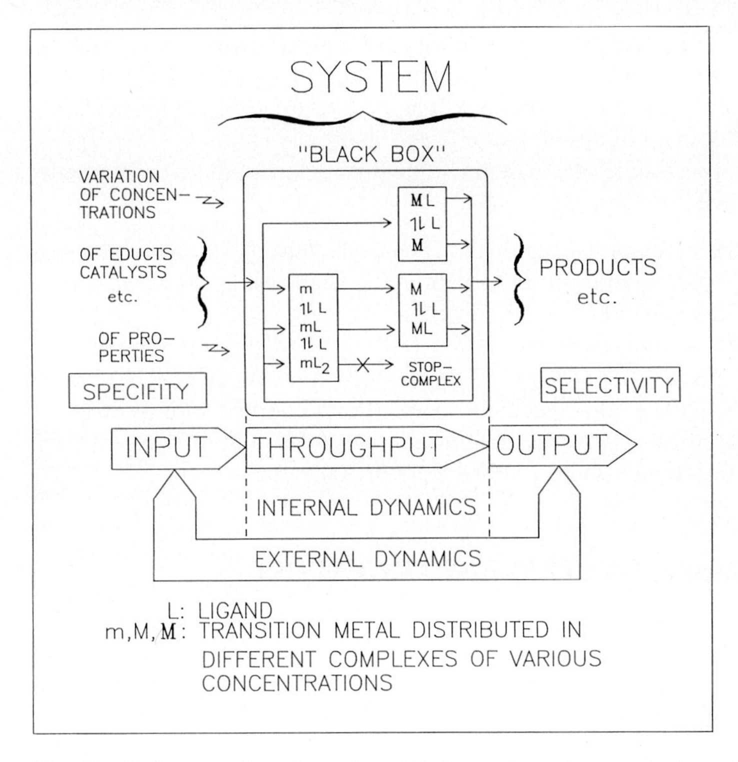

Fig. 7.1. Definitions for a ligand modified metal catalyst system (see also text)

what is still more striking – is moreover *exact only for the system at the moment under investigation*. Very small variations in only one compound or partial component of the INPUT (property change of one educt, solvent or catalyst e.g. only by variation of a single atom or single substituent in these compounds) can dramatically change the couplings in the "black box" (see examples in Chaps. 3 to 5). This is normally overlooked or not mentioned in the "mechanistic" text books. Therefore we need in addition to and specially before such careful investigations a second strategy to elucidate qualitatively the whole system.

The so-called external dynamics describe merely INPUT-OUTPUT correlations. Only systematic variations of the initial conditions of the INPUT, e.g. by changing concentrations (Chap. 6) and varying alternative properties of the educts, solvents and catalysts are considered to have direct consequences in their relations to changes of product distribution. A detailed analysis may lead, by means of model concepts, to a more or less defined picture of the "black box". But very often contradictory models can correctly express the INPUT-OUTPUT correlations. Nevertheless we are convinced, that systematic INPUT/OUTPUT investigations are the fastest procedure for an evolution of a complex chemical system. Exact internal dynamics, which will *not* be discussed in this monograph, should be done after evolution of the desired chemical process. Our ordering CONCEPT based on ALTERNATIVE and alternating PRINCIPLES (see Chaps. 3 to 5 and 8) is especially worthwhile as a model for general phase relation control of development of alternative order in chemistry.

With Fig. 7.1 we would like to define some further terms, that are frequently used in a controversial way, e.g.:

Selectivities (stereo-, enantio-, regio- and chemoselectivity [172] and so on) – by our definition – only give information about distributions in the reservoirs of the products of the OUTPUT.

The corresponding *specificities* of the INPUT show the relations of their ALTERNATIVES to those of the OUTPUT (vide Chap. 8). ALTERNATIVES of the THROUGHPUT (also specificities) are always coupled with those of the INPUT [23]. Specificities – by our definition – are not bound to have anything in common with statements as e.g. 100% yield or 100% purity.

7.2 Intermolecular SYSTEM ENLARGEMENT

An elegant solution of the problem of how to influence reactivities is offered by intermolecular SYSTEM ENLARGEMENT. Thereby we can achieve the desired or surprising new reactivity of the original basic system by addition of one or more components. The new compounds are able either to enhance or to reduce the rate of a given reaction or to open up new pathways of reactivity. The necessary experimental method to determine the correct amount was described in detail in Chap. 6.

In this context we want to point out what seems to be an inconsistency (a contradiction). If you add more and more compounds in increasing quantities to a pure substance, this substance, of course, will become increasingly contaminated. But if you add to the educts of a chemical process, the correct quantities of catalysts and additional effectors with suitable qualities, the catalytic process becomes more and more "purified", so that in certain (well-suited) cases only one single product is formed very rapidly. Mittasch stated: "The multi-component catalyst is the winning catalyst!" [218]. By skilfull intermolecular SYSTEM ENLARGEMENT the qualitative composition of the products in the desired process may remain unchanged and there arise only quantitative changes in individual rates or in the overall rate. But it is normal to enlarge and vary the possibilities of syntheses greatly by intermolecular SYSTEM ENLARGEMENT.

In Fig. 7.2 the product spectrum is extended by enlargement of a two-components-system to a five-components-system. By marking, we have pointed out e.g. the selective formation of two different structural isomers in two five-component-systems only by changing one ALTERNATIVE PRINCIPLE (see Chap. 8) within the 2:1-cooligomers of butadiene and Schiff-bases. The original four-components-system $Ni^{(0)}$/butadiene/Schiff-base/TPP produces both

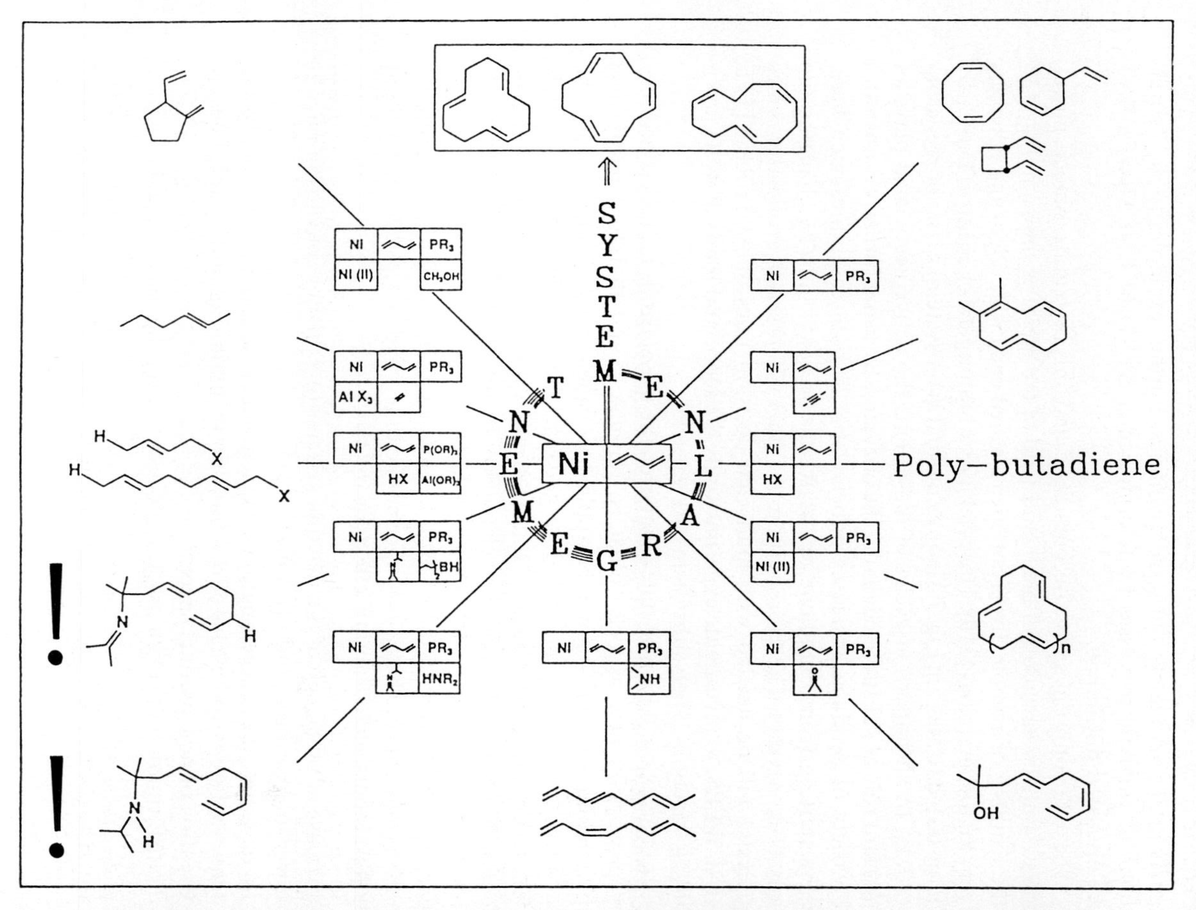

Fig. 7.2. Widening of product spectrum by SYSTEM ENLARGEMENT of the two-components system $Ni^{(0)}$/butadiene up to five-components systems

isomers in approximately the same quantities. Further examples to the "purification" of catalytic processes by intermolecular SYSTEM ENLARGEMENT were discussed in Chap. 6.

The use of only one substrate, butadiene, and only one catalyst-metal, nickel, allows – only by variation of the ligand-field of the metal as shown in Fig. 7.3b – the synthesis of a whole display of openchained and cyclic oligo- and polymers in very high selectivities [219]. This is a kind of biomimetic chemistry. The partly demonstrated strategy in Fig. 7.3 is quite normal in biosystems as can be seen from Fig. 7.3a. In all important cycles of biosystems the active acetic acid plays an important rôle as substrate changing the enzymes. By using further cosubstrates the variety of products increases unbelievably in both cases presented here, always starting from butadiene or "active acetic acid" only by changing the catalysts or enzymes. Therefore perhaps in future whole sets of artificial enzymes (catalysts) will be needed with multidual decision trees of ALTERNATIVE PRINCIPLES, to investigate experimentally the scope and limits of well known catalytic reactions when subsystems are varied.

In conclusion, it may be stated that even the use of catalysts does not convey any principally new aspects. What is unusual – and therefore partly new experimental strategies have to be applied to some extent – is the fact that the systems which are to be examined in the "black box" are present in extremely small steady state concentrations, thus elude direct analysis. But also in catalytic processes we have to suppose that changes of quality are induced only by newly arising influences (ALTERNATIVE PRINCIPLES). Therefore, we have to regard the catalysts with their partly hidden associations with educts, interim products and intermediates as "supermobile functional groups", bringing about amongst others such dramatic effects as follows:

1. Increase of efficiency of the reaction systems (e.g. higher conversion rates at lower temperatures and pressures).
2. Increase of the exceedingly various product selectivities by activating requested steps and cycles of reaction and by inhibiting undesirable ones.
3. Creation of new product spectra.
4. Transformation of stoichiometric processes into catalytic ones and so on.

→

Fig. 7.3. a The multifold application of the substrate "active acetic acid" in different cycles with special enzymes in biosystems is in (**b**) compared with a type of biomimetic chemistry, wherebyt the individual products are formed out of the substrate butadiene at the catalyst-metal Ni alone by variations of the ligand field in different cycles (modes)

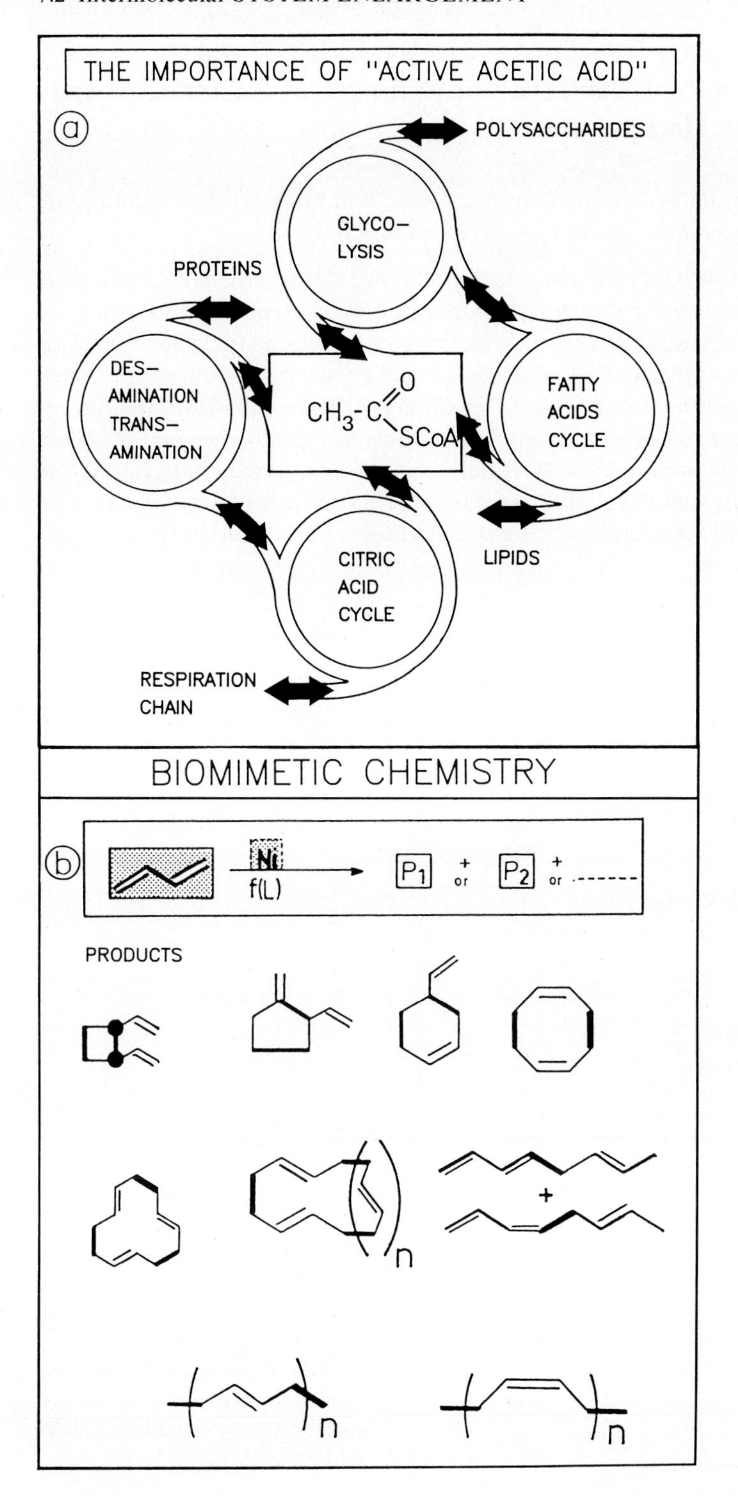
THE IMPORTANCE OF "ACTIVE ACETIC ACID"
a
POLYSACCHARIDES
GLYCO-
LYSIS
PROTEINS
DES-
AMINATION
TRANS-
AMINATION
CH3-C=O SCoA
FATTY
ACIDS
CYCLE
LIPIDS
CITRIC
ACID
CYCLE
RESPIRATION
CHAIN
BIOMIMETIC CHEMISTRY
b
Ni
f(L)
P1
P2
+ or
PRODUCTS
n

7.3 Intramolecular SYSTEM ENLARGEMENT and VARIATION: Substitution of Hydrogen by Substituents and Carbon by Heteroatoms

When we substitute hydrogen atoms in a parent-compound (Fig. 7.4a and b) we have to distinguish two different situations precisely.

(1) The new substituents introduced only influence a determined reactivity of the parent-compound under observation, caused by quantitative variations of the individual rates of reaction and, thereby, of the product spectrum. This case and its limitations were discussed more thoroughly in Chap. 2, when arguing about the limitations of *F*ree *E*nergy (FE-) and *L*inear *F*ree *E*nergy (LFE-) relations. System variations as shown in Fig. 7.4b are performed by following the well-known "Grimm'schen Hydridverschiebungssatz" (see Fig. 7.5). The ordering influence of these replacements of heteroatoms with an EVEN/ODD number of valence electrons was described by the examples in Fig. 2.14–18.

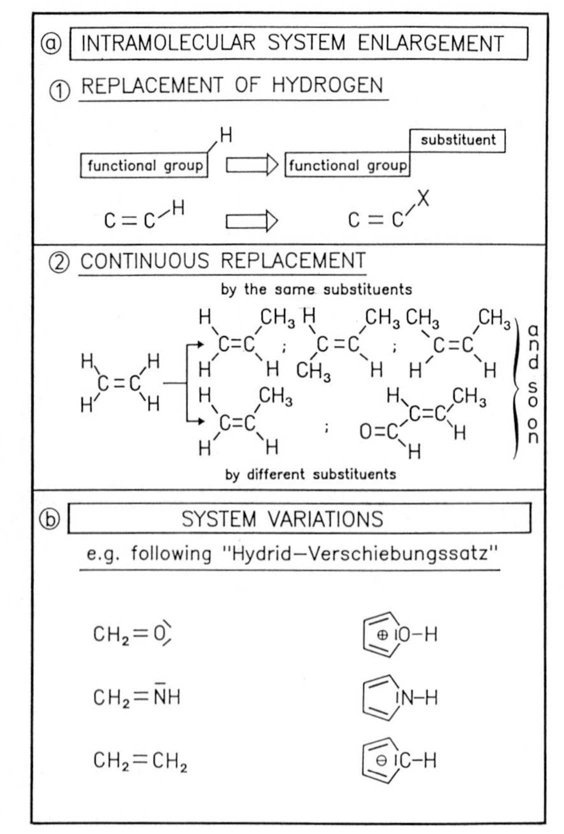

Fig. 7.4a,b. The intramolecular SYSTEM ENLARGEMENT (variation *at* the system) in **(a)** by introduction of substituents (classes of atoms) instead of hydrogen is compared in **(b)** with the variation *in* the system by ATOM REPLACEMENT

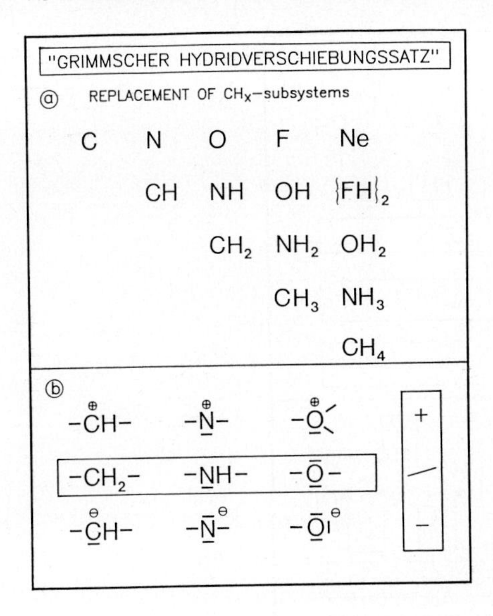

Fig. 7.5a,b. The "Grimm'sche Hydrid-Verschiebungssatz" in neutral **(a)** and charged **(b)** subsystems

(2) When, by adding new substituents, these subsystems become effective at the same time as new functional groups, qualitative changes of the product spectrum will arise (new possibilities of reaction appear). We want to deal briefly with the second case. By the example of the direct unifying of C-X- or alternatively C=X-groupings with C-H-acid functions (see Chap. 3.5) we would like to point out in Fig. 7.6, how the number of the potential syntheses came into existence by the added substituent now active as a new functional group. In this case even alkyl groups as substituents will act as functional groups (C-H-acid compounds).

The method, shown in Fig. 7.6, of demonstrating the relations in chemical systems of increasing complexity should be adopted by the chemist as soon as possible in order to familiarize himself with the couplings of many individual information (associative learning in patterns).

An important aspect in the introduction of controlling substituents is the question of how easily these groupings as mobile functional groups can be eliminated again from the products by very simple reactions (hydrolysis, hydrogenation etc.). One of the new possibilities of transferring a mobile group arousing much interest is being used for the synthesis of polymers in the so called group-transfer-polymerization [220] (vide Fig. 8.8 later on). Of outstanding elegance, of course, are those examples, where the substituents or subsystems introduced in the basic system as pertubation, are split off – so to say, automatically in the course of the reactions – after having accomplished their task of varying the reactivity (see examples of Wittig-, elimination- and analogous reactions in Chap. 5). Under these aspects catalysts are "supermobile functional groups" (self-addition and -elimination via equilibria).

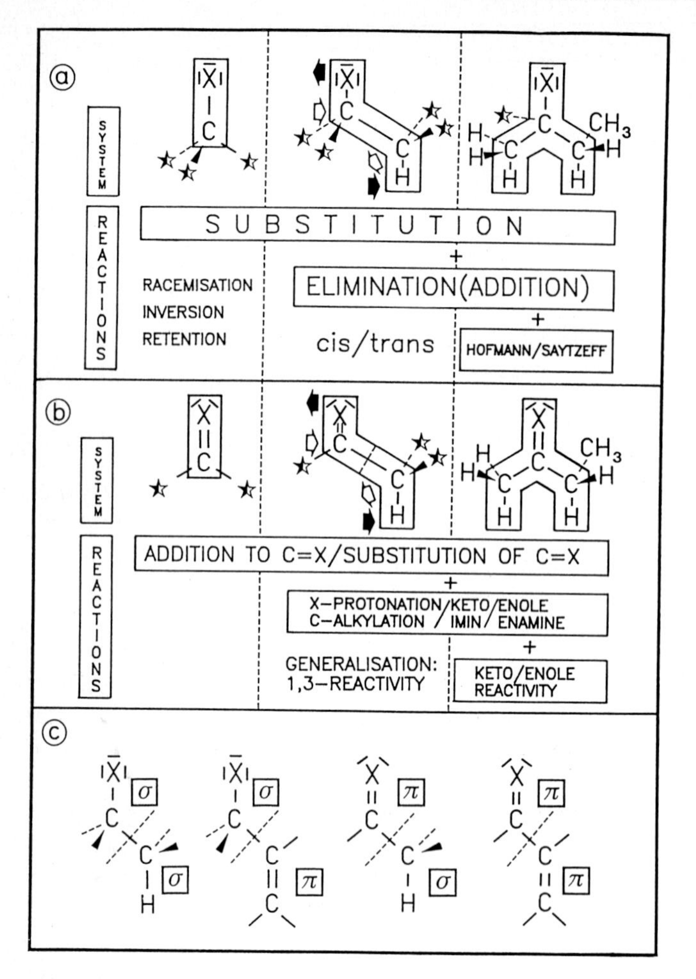

Fig. 7.6a-c. A direct unifying of "functional groups" (C-X **(a)** or C = X **(b)** with C-H respectively $C_{pr.}$-H/$C_{sec.}$-H) leads in a related manner (pattern that connects [99] to an enlargement in the possibilities of reaction types and their control. **(c)** shows further combinations

7.4 Coupling of Subsystems

We would like to present a far more abstract application of inter- and intramolecular SYSTEM ENLARGEMENT by general unifying principles. The formal procedure in perturbation theory [41] was demonstrated step by step for two ethenes and two allylradicals in Fig. 3.15a. This is a theoretical abstract procedure and cannot be verified directly by experiments. But unifying via a metal as was also shown in Fig. 3.15b can be realized as a model as well as a direct procedure in reality.

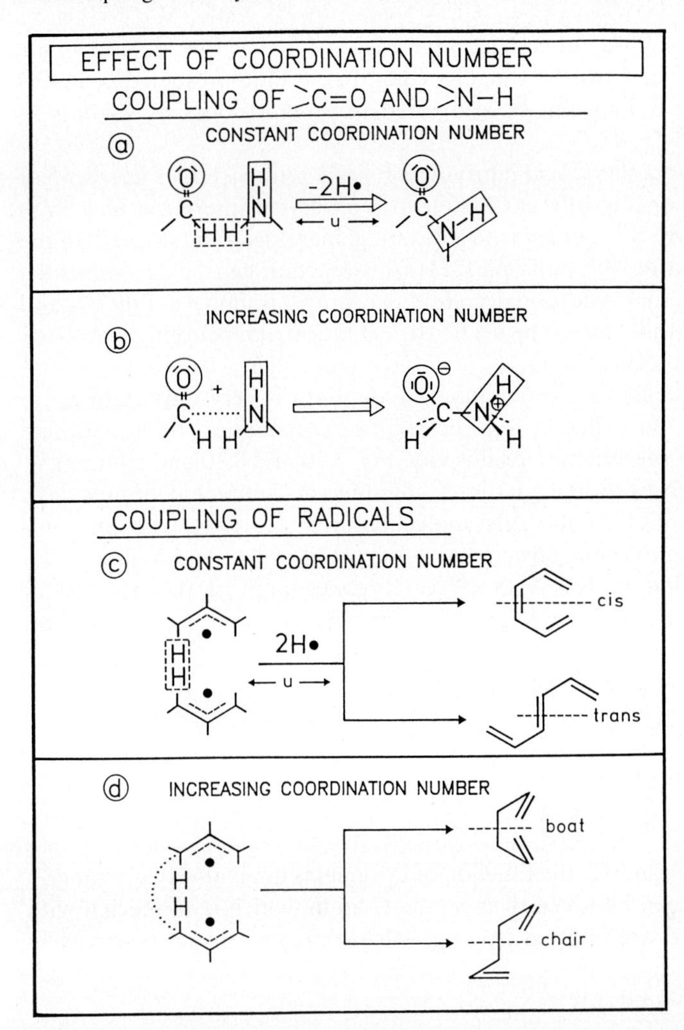

Fig. 7.7a-d. The dramatic difference between a direct unifying (formally H-abstraction and coupling of radicals) under constant coordination number ((**a**) and (**c**)) and addition (real reaction) with increasing coordination number ((**b**) and (**d**)) is demonstrated by two examples

In order to demonstrate, how differently intra- and intermolecular SYSTEM ENLARGEMENT and VARIATION may operate, we shall compare the direct unifying of two functional groups (C = O and N–H, where the C = O-group is arbitrarily defined as a parent-system) first maintaining up and then by increasing (Fig. 7.7) the coordination number at the centers of coupling.

After having eliminated one H-radical in the parent-system and one in the enlargement component, the direct unifying of both residues represents formally an intramolecular SYSTEM ENLARGEMENT. Thereby, the coordination number is maintained, but, of course, there ensues a noticeable decrease in the strength of the base and the N-H-acidity (the process here presented is merely imaginary, of course, and can be converted into reality only by detours in experiments).

On the contrary, the direct unifying of both groups under increase of coordination-number is a different matter. It is a realistic intermolecular SYSTEM ENLARGEMENT and leads to a dramatic increase of the strength of the acid and the base after the unifying [221] (this reaction can be accomplished directly by experiments). Additionally, another example is shown in Fig. 7.7c and d (here the systems that have to be unified are identical and coupling is realized with alternative stereochemistry).

The inverse process to direct unifying (coupling) is very important as an abstract procedure, too, called by us direct splitting (decoupling) or the coupling of both imaginary experiments (for this vide Fig. 3.20 and 8.10 and context).

Besides the substitution and direct coupling of functional groups in a molecule the concept of the so-called molecular architecture [45], i.e. the composition of complex molecular structures out of partial structures, has repeatedly proved effective for decades. A first great breakthrough of this method of reasoning can be certainly seen in the so-called vinylogy- and phenylogy-principle [222].

If in a compound of the type

$$A - E_1 = E_2 \text{ or } A - E_1 \equiv E_2$$

a structural unit of the type

$$-(CH = CH)_n - \text{ or } \left(\text{C}_6\text{H}_4\right)_n$$

is inserted between A and E_1, the function of E_2 remains qualitatively unchanged. The function of E_1 can be taken over by the C atom, which is connected with A. Symmetry aspects are omitted (see e.g. Michael/Knoevenagel reactivity in Sect. 5.2).

Resulting compounds are:

$$A - (CH = CH)_n - E_1 = E_2 \text{ or } A - (CH = CH)_n - E_1 \equiv E_2$$

We have already given a more abstract application of intra- and intermolecular SYSTEM ENLARGEMENT and VARIATION by means of a metal-catalytic system by introducing the so-called METALALOGY PRINCIPLE [39] (vide Fig. 3.17 and 18). For precise and creative thinking in system theory it is necessary to separate the invariant parts (solid lines) and the variable parts (broken lines) of the whole system. In reality this is not possible (Moiré-phenomenon).

7.5 Symmetry Aspects in the Coupling of Chemical Subsystems

The so-called Woodward-Hoffmann rules [48] e.g. the type of ring closure (con-/dis-rotatory) in alternating vinylogous systems, are represented by absolute formulations (forbidden/allowed). These rules applied in relative form (stabilized/destabilized and activated/inhibited) lead to the multidual decision trees, already introduced in Chap. 5. A detailed comparison of absolute and relative symmetry rules is described by examples in [23] and will not be repeated here. Instead of this we will demonstrate by another example, that a cascade of ring closure reactions in two directly unified 1,3-dienes – they correspond then to conjugated *n*-tetraenes – and in two dienes coupled via a catalyst (see the comparison of direct unifying and metalalogy principle amongst others in Fig. 3.14) leads to comparable stereochemistry in the final products. As a first direct hint to the validity of this rule, in our opinion, is the perfect equivalency in two consecutive ring closure reactions in planar π-systems [223,224] and in analogous metala ring closure reactions [150,225], which show the same substitution pattern in the formed four-membered rings (vide Fig. 7.8). The direct compatibility is by chance and by exact inverse symmetry of coupling by choosing catalysts with ALTERNATIVE PRINCIPLES (e.g. Ti- or Pd-catalysts) the alternative cascade should be realized.

One aspect has to be considered in more detail. We were able to prove the existing regularities of the metala ring closure experimentally only by applying the principle of microscopic reversibility. The *trans, cis*-3,4-dimethyl-*cis,cis*-1,2-divinylcyclobutane formed out of piperylene is split at $Ni^{(0)}$-L-catalysts to *trans*-piperylenes under reduced pressure (see also the equilibria of 1,3-dienes with the corresponding four-membered rings in Fig. 3.20a), which are immediately separated from the reaction mixture by evaporation. This is more

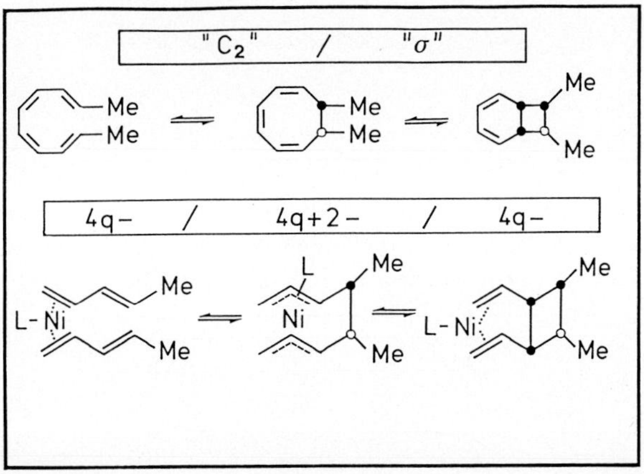

Fig. 7.8. Comparison of stereochemistry of ring closure reactions of direct unified π-systems with metala ring closures

important in the case of the *trans,trans*-3,4-dimethyl-*cis,cis*-divinylcyclobutane, which is split at the catalyst to one *trans*- and one *cis*-piperylene.

The *cis*-piperylene is rapidly isomerized at the same catalyst to *trans*-piperylene [150]. These processes with low activation energies, which invert local symmetry, obey strong symmetry laws, too, (See the OCAMS-method of Halevi [226]). It is perhaps better to speak of local symmetry inversion instead of a "symmetry breaking process". The assumption often found, that a reaction shows only X% concertedness, masks the fact, that the symmetry rules are always valid – though often only relativated – and that there are symmetry inverting processes with small activation energies as shown in an abstract form in Scheme 7.1 (as we are familiar with changes to alternative conformations). So finally e.g. a *cis-/trans*-isomerization of a double bond may be realized by conformational changes using addition/elimination reactions of low activation energy (for this vide Fig. 10.3).

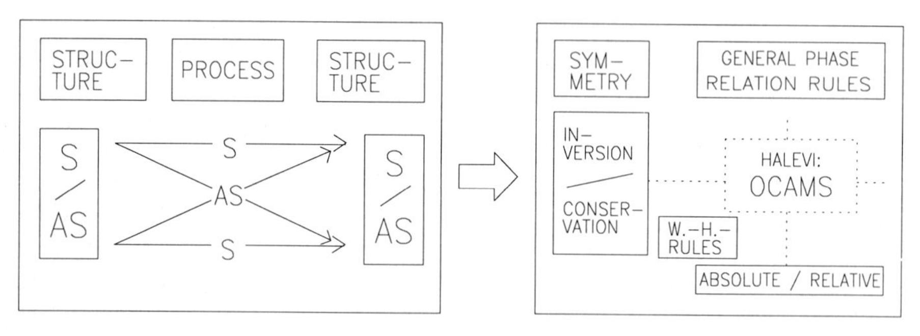

Scheme 7.1

8 Models and Methods for the Understanding of Self-organization and Synergetics in Chemical Systems

"Es ist hier nicht die Rede von einer durchzusetzenden Meinung, sondern von einer mitzuteilenden Methode, deren sich ein jeder als eines Werkzeugs nach seiner Art bedienen möge."

J.W. v. Goethe, Farbenlehre II

Whenever information is transferred by language one differentiates between syntactic, semantic and pragmatic aspects. The transferring of information about the facts of a deeper understanding of self-organization and synergetics of chemical systems in our opinion finds its analogy to this in the following statements. The reader will find the formalized and mathematized – and thereby subject separated – objective descriptions (the syntax) in physics, physico-organic and theoretical chemistry. This will be briefly mentioned in Chap. 10, as motivation for further studies.

The pragmatic aspects of self-organization – sometimes controlled by very simple ALTERNATIVES – may be identified most impressively in the highly evolved and organized chemical systems and subsystems of the biosciences (see hints given in Chap. 9). In the simple, but still not evolved systems whose development the experimentalist involved in synthesis has to take care of, the meaning and controlling power (the semantics) of individual ALTERNATIVE PRINCIPLES become especially evident.

Two corresponding double – dual procedures are presented and described in an abstract manner in Fig. 8.1 and 2 to find a more or less consistent form for the CONCEPT OF ALTERNATIVE PRINCIPLES. The starting points differ totally for both types of the unifying description. While the Aristotelic idea of exploring thoroughly the four roots of argumentation for a possible single deep truth gives the background to the formalism presented in Fig. 8.1, the point of departure for the pathways of elaboration in Fig. 8.2 giving the present state of knowledge in chemistry. In both cases our common aim is to produce a representation and an aid to comprehension of the connecting patterns (9).

8.1 The Ordering CONCEPT OF ALTERNATIVE PRINCIPLES in a Comprehensive Form

In Fig. 8.1, two causes in each case are offered for going upwards and downwards in the levels of chemistry. The electrons (e.g. the number in the valence shell, see Fig. 2.14), the atoms (e.g. described by their relative first ionisation potential

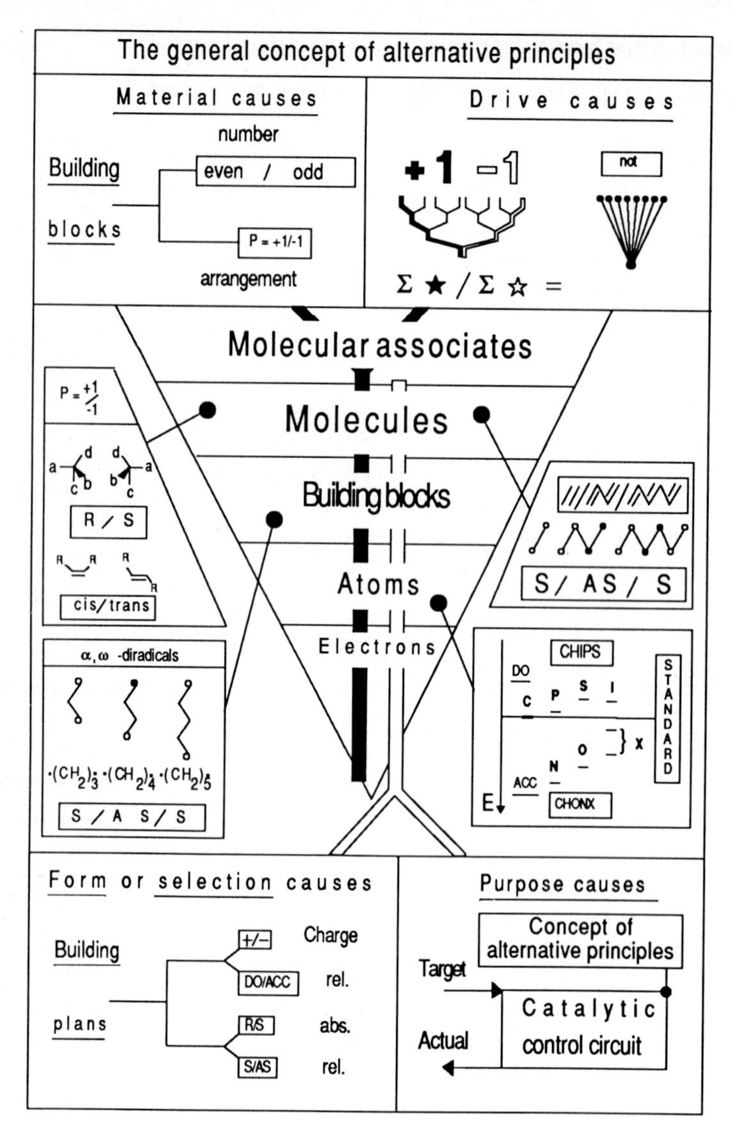

Fig. 8.1. The four good reasons (causa materialis, efficiens, formalis and finalis) – of one perhaps uniform, deep truth – are presented in schemes and examples by an ordering, general CONCEPT OF ALTERNATIVE and alternating PRINCIPLES in systems of increasing complexity (see text)

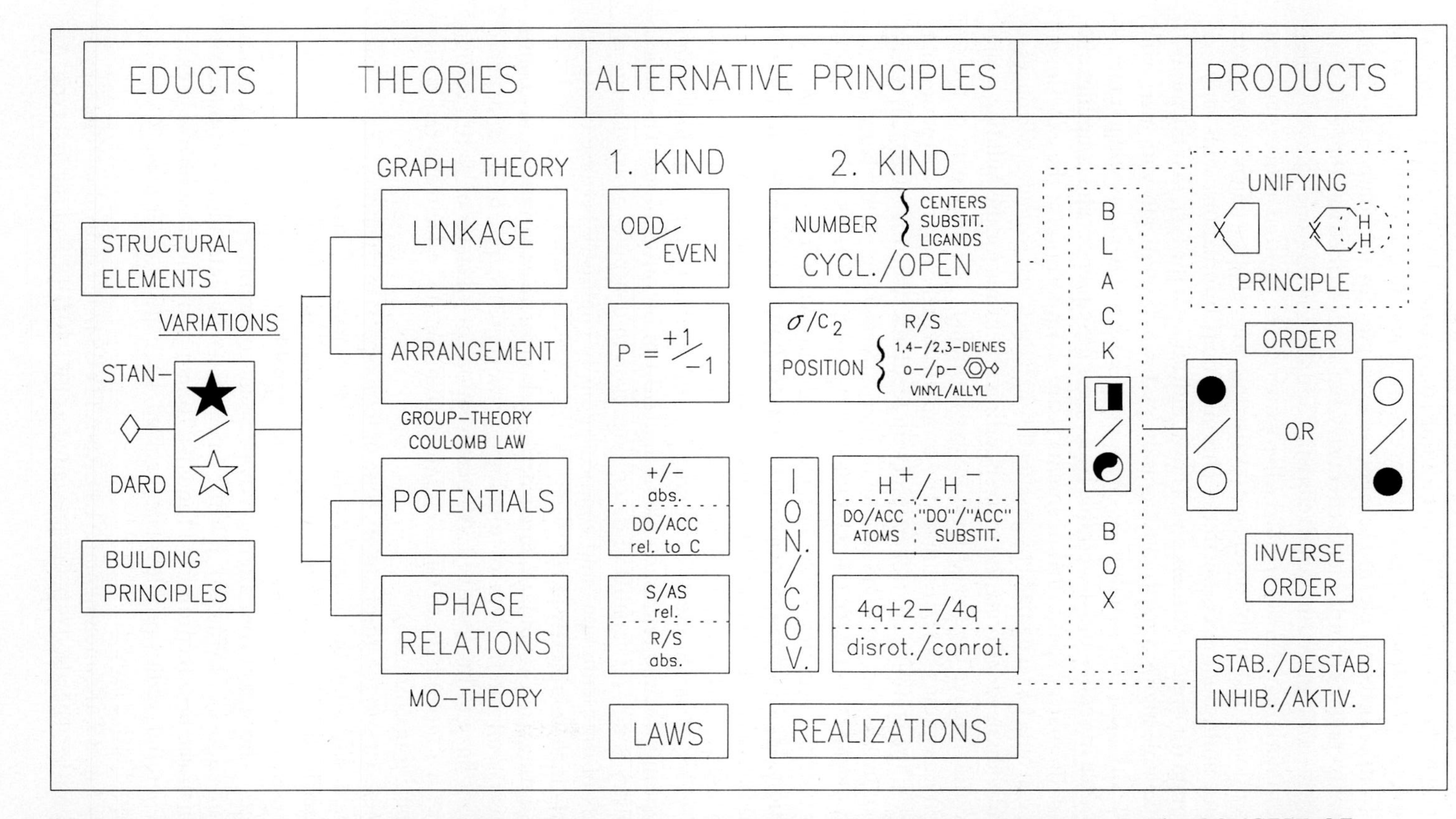

Fig. 8.2. Starting from the fact, that the alternatives in chemistry are describable by four partial aspects, the CONCEPT OF ALTERNATIVE PRINCIPLES is presented in an other mode of description. A separation of original (1. type) and heuristic, derived ALTERNATIVE PRINCIPLES (2. type) is proposed

based on carbon as standard), the building stones (their EVEN/ODD number), the molecules (with their spatial arrangement or alternative symmetry S/AS of their frontier orbitals) etc. are selected as examples for the components of the levels of increasing complexity in chemistry and are described together with some alternative and alternating characteristics and criteria. On one side only alternative material causes (causa materialis), which can be realized in an experiment, are mentioned and on the other side the driving causes (causa efficiens) are given forming multi-dual decision trees. The multidual bifurcations with only one maximum for opposite orders give a clear indication of the fact, that the number of achievable highly organized systems drops down dramatically with increasing complexity after a maximum [8]. Why these EVEN/ODD building stones – applicable in reality and discernible by number of sequences and couplings and type of alternative spatial arrangements – are so important for the experimentalist, can only be recognized by passing through the levels from the collectives to the individuals identifying the corresponding blueprints for the building stones in the next higher organized level (causa formalis). The effectivity of the organization systems can be estimated by comparison of the nominal/experimental values in the complex control circuit (with feedback) (see Fig. 10.2) [23] (causa finalis). In a perfect catalytic system all alternatives are in perfect harmony.

With the CONCEPT OF ALTERNATIVE PRINCIPLES we are trying to generalize the symmetry rules, which are already known in an absolute formulation [48], by means of phase relation rules (vide Fig. 10.2a) in the form of multi-dual decision trees for alternative order. Accepting the cooperativities and compensations in pairs (as phase relations: IN + IN, OUT + OUT, IN + OUT, OUT + IN) of small effects as alternatives of symmetry relations makes this concept in principle representable in digital form by bifurcations. Theoretical chemistry is not yet capable of describing, even in computer-assisted form, all aspects of differentiations and compensations in chemistry because of the extremely high complexity [227].

Starting from a standard, surely being carbon for an experimenter in organic chemistry, all alternatives of topology, arrangements, potentials and phase relations have to be described. All perturbations of all states of $\approx 10^{23}$ carbon atoms (in a mole) have to be taken into account in the collective to rationalize chemical transformation.

Referring to the graph theory (one of four aspects in chemistry) the changes can be described in the form of matrices [228]. Normal group theory is not able to describe the changes from one symmetry class to others. Nevertheless new strategies are available for overcoming these problems – in the second aspect of changes – brought to our attention by personal communications from Altmann (Oxford) and Ugi (Munich). For the electromagnetic interactions in materials (not in molecules) according to Primas [227] changes are not describable in exact terms yet. Before this problem is not solved, the experimenter must be content with qualitative models.

In Fig. 8.2 all fundamental ALTERNATIVES in the individual four aspects of chemistry (see also [74]), are presented as PRINCIPLES of the first type. All such ALTERNATIVES possessing a high heuristic value and being derives from those of the first type, are termed PRINCIPLES of the second type. But they are just these ALTERNATIVES which as known realizations make this concept more attractive for pragmatic experimenters.

We pointed out the problems in the coupling of paritetic (symbol: ◨) and complementary (symbol: ◐) ALTERNATIVE PRINCIPLES (in the THROUGHPUT) in Sect. 5.3.

8.2 Some Statements to the Application of the CONCEPT OF ALTERNATIVE PRINCIPLES

Fig. 8.3 shows by means of examples, some differentiations in chemistry which are commonly described as "problems of three"; but those problems can be solved experimentally with the help of the CONCEPT OF ALTERNATIVE PRINCIPLES only, if the hidden double-dual differentiation is clearly recognized.

The unequivocal description in (a) concerning a conformational or configurational problem is widespread. In the discussion of e.g. configurations in pentadienyl-anions or 1,3-disubstituted allylic anions only three possible arrangements given as W-, S- and U-forms are sometimes described (see (b)). Considering phase-relations in π-systems, too, one should describe e.g. the forms M/U and W/"roof" as alternatives of e.g. highest and lowest stability.

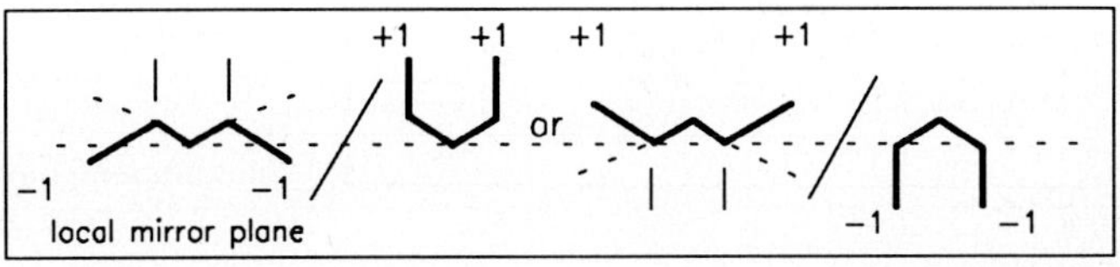

It is well known from propene-dimerization (see (c)) [203], that four pathways to the three skeletons of the dimeric hydrocarbons exist. In Chap. 9 we will mention briefly in context with Scheme 9.1 the problem, that highly evolved systems really give rise to only three solutions (but in 1:2:1 ratio).

We have already accentuated that (see Chap. 5) the hierarchical order of ALTERNATIVE PRINCIPLES is system dependent. By the examples summarized in Fig. 8.4 [229–231] we would like to emphasize, that the hierarchical order of energy contributions to the stabilization of alternative structures may be changed and is system dependent, too.

At the same time, we would like to demonstrate by the examples in (a) and (b), that alternative stereochemical substitutions in a four-membered ring – far away from the reaction center of a Cope rearrangement – strongly influence the populations of stereochemically differentiated pathways of the whole process. An alternative *cis-/trans*-disubstitution by methyl groups in 3,4-positions controls the population of so-called six-or four center rearrangement [150,230], while the parent system *cis*-1,2-divinylcyclobutane – only substituted by hydrogen atoms

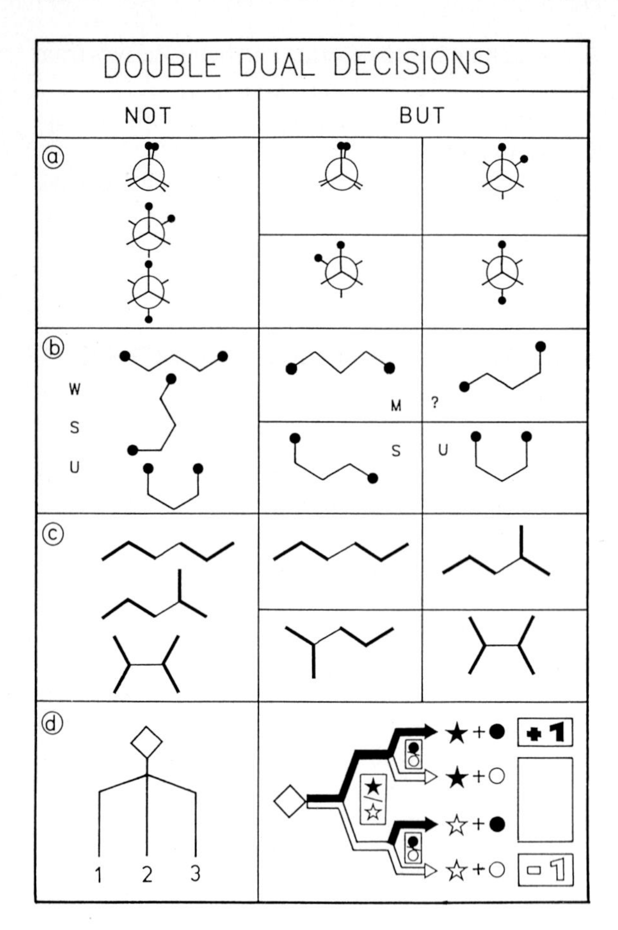

Fig. 8.3a-d. A lot of differentiations in chemistry are based on double-dual decisions, but not on "threefold differentiations"

in these positions – is transformed at 40°C in a ratio of 97:3 to *cis,cis*-1,5-cyclooctadiene or the highly strained *cis,trans*-1,5-cyclooctadiene [229].

The direction of the Cope rearrangement is strictly determined in every case going from the four- to the eight-membered ring both by ring strain (four-membered > eight-membered ring) and the degree of alkyl substitution at the double bond (in the four-membered ring one and in the eight-membered ring two alkyls per double bond). The difference of the energies arising from the ring strain (ten-membered > six-membered ring) exceeds the energy value of the oppositely directed effect of lower alkyl substitution in the six-membered ring (one alkyl group per double bond) compared to the ten-membered ring (two alkyls per double bond). The energy difference in ring strain is in a higher position in the hierarchical order than the degree of alkylation (Fig. 8.4c).

The hierarchical order of these energy differences can change depending on the system, so that ten-membered rings are formed from six-membered ones by rearrangement [230] (compare Fig. 8.4c and d).

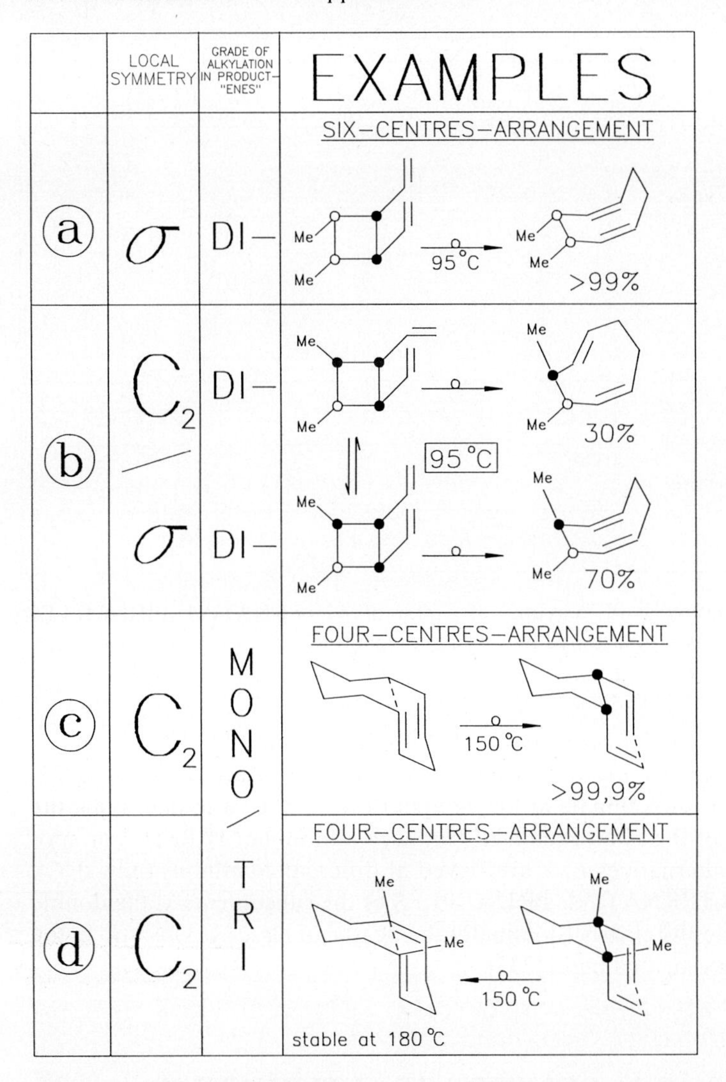

Fig. 8.4a-d. The preference for one order in the stereochemistry of Cope rearrangement in (**a**) disappears changing to an alternative *trans*-arrangement of both methyl groups in 3,4-positions of the four-membered ring in (**b**). While the direction of the Cope rearrangements in (**a**) and (**b**) is unequivocally determined both by ring strain and the degree of alkylation at the double bond, a change in the hierarchical order of the two opposing effects here causes a change in the direction of the reactions of (**c**) and (**d**)

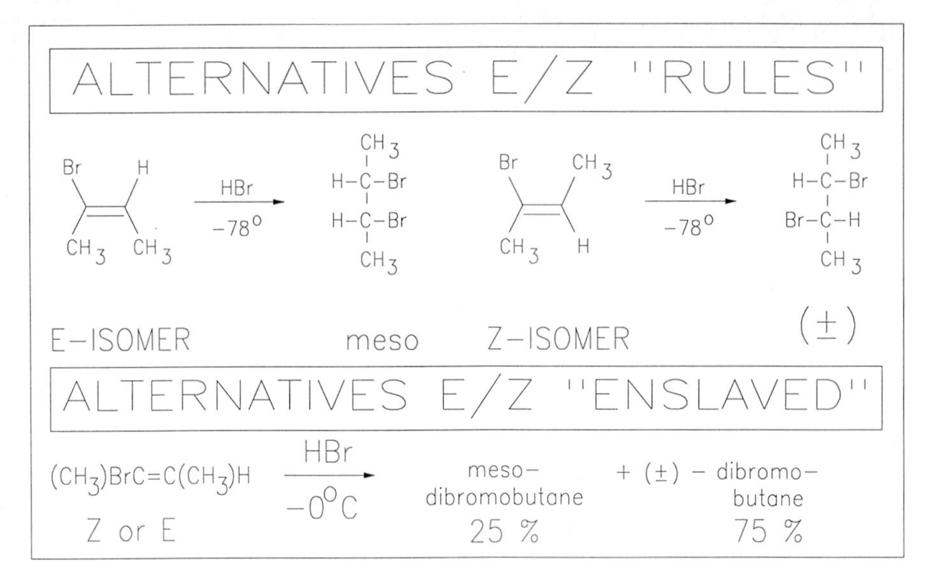

Fig. 8.5. "Dominating" and "slaving" of a pair of ALTERNATIVE PRINCIPLES (*cis-/trans* or *E/Z*) is demonstrated by varying temperature

The same is true for pairs of ALTERNATIVE PRINCIPLES as demonstrated by an example in Fig. 8.5. Referring to the nomenclature of Haken [1,232] the alternatives *E/Z* dominate at low temperature (–78 °C) and determine the stereochemistry of the photochemically induced addition of H-Br to 2-bromo-but-2-ene. The alternatives *E/Z* are slaved at different conditions (now 0 °C). Now e.g. the ALTERNATIVE PRINCIPLES of the substituents at the double bond determine quite clearly (dominate) the degree of *meso/racem*-ratio in the products (not investigated here) [233].

8.3 Application of the CONCEPT OF ALTERNATIVE PRINCIPLES as Compensation Strategy

In Fig. 8.6, some methods and strategies are summarized, those which we applied in the course of developing the ordering concept. As we did so, it becomes obvious, that – as described in the preceeding Chaps – a few new aspects of the long-standing methods appeared. Now we would like to emphasize the advantages of compensation phenomena, because the information destroying aspects of these phenomena were featured strongly in Chap. 2. In Fig. 8.7, a very impressive example is given, that an inhibition of a reaction cascade (replacement of a substituent – OCH_3 by –SPh in one subsystem) can be balanced by skilful compensations in many subsystems. The corresponding alternatives following

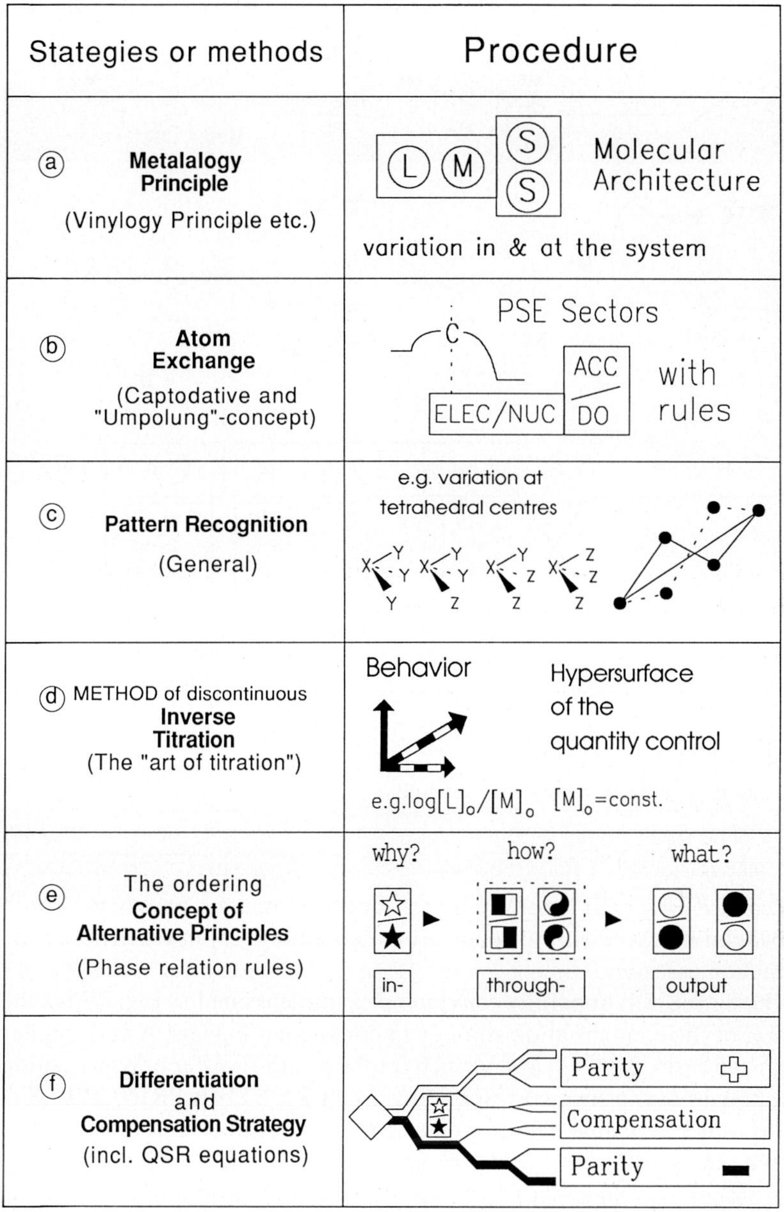

Stategies or methods	Procedure
ⓐ **Metalalogy Principle** (Vinylogy Principle etc.)	Ⓛ Ⓜ Ⓢ Ⓢ Molecular Architecture; variation in & at the system
ⓑ **Atom Exchange** (Captodative and "Umpolung"-concept)	PSE Sectors; C; ELEC/NUC; ACC/DO; with rules
ⓒ **Pattern Recognition** (General)	e.g. variation at tetrahedral centres; X Y Y Y; X Y Y Z; X Y Z Z; X Z Z Z
ⓓ METHOD of discontinuous **Inverse Titration** (The "art of titration")	Behavior; Hypersurface of the quantity control; e.g. $\log[L]_0/[M]_0$ $[M]_0$=const.
ⓔ The ordering **Concept of Alternative Principles** (Phase relation rules)	why? how? what?; in- through- output
ⓕ **Differentiation** and **Compensation Strategy** (incl. QSR equations)	Parity ✚; Compensation; Parity ▬

Fig. 8.6a-f. Those strategies and experimental methods are summarized, which were often successfully applied for recognizing the rôle of ALTERNATIVE PRINCIPLES in the self-organization of chemical systems.

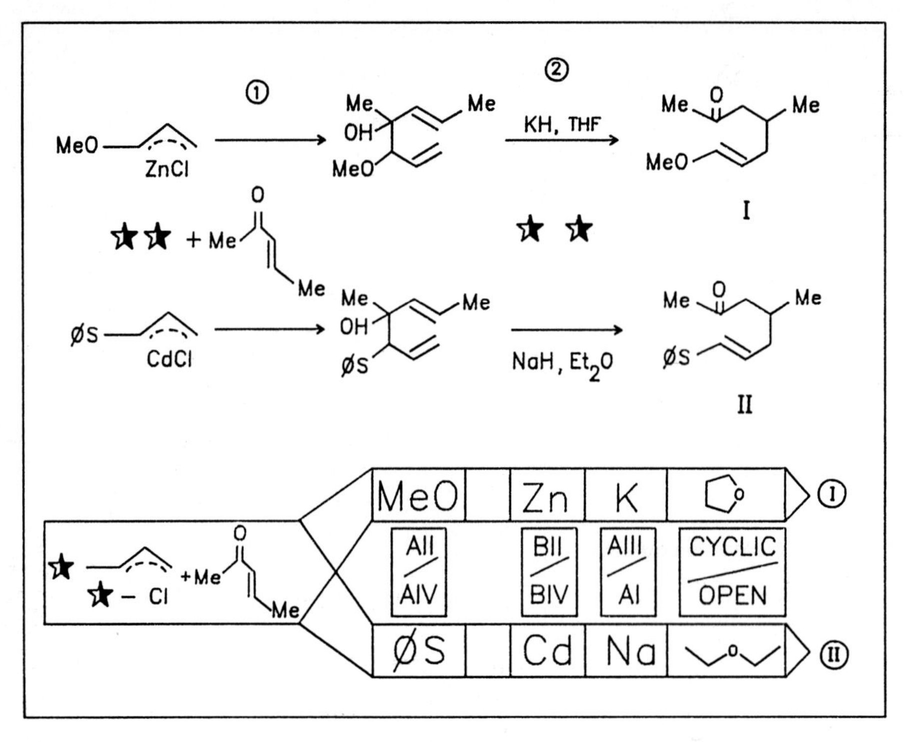

Fig. 8.7. The replacement of only one substituent, $-OCH_3$ versus –SPh, makes a change to ALTERNATIVE PRINCIPLES in subsystems necessary to compensate for the perturbation and to reactivate the reaction cascade

our concept are presented here. Without this concept one has merely to admire the overwhelming experience displayed by the authors [15] in reactivating the whole reaction cascade.

A very simple, but hopefully convincing example is cited in Fig. 8.8 for the application of the compensation strategy to circumvent patents. A very similar effector in two cases [220,234] is described for the group-transfer polymerization by the compensation of two corresponding ALTERNATIVE PRINCIPLES – namely 4q–/4q + 2-π-subsystem as well as DO/ACC of the heteroatoms in this part keeping the substituent – $SiMe_3$ constant.

"The preparative challenge to perform stereoselective synthesis has one of its origins in the observation of tremendous pharmacological and physiological differences in the effects of enantiomeric compounds (vide Fig. 8.9). Since the thalidomide catastrophy all those engaged in this business are conscious, how necessary it is to apply stereochemically homogeneous, i.e. enantiomerically pure, drugs. However, striving with the single aim of developing enantioselective synthesis, can hide a further important aspect, which Knabe and Schamber demonstrated impressively by an example from pharmaceutical research. A

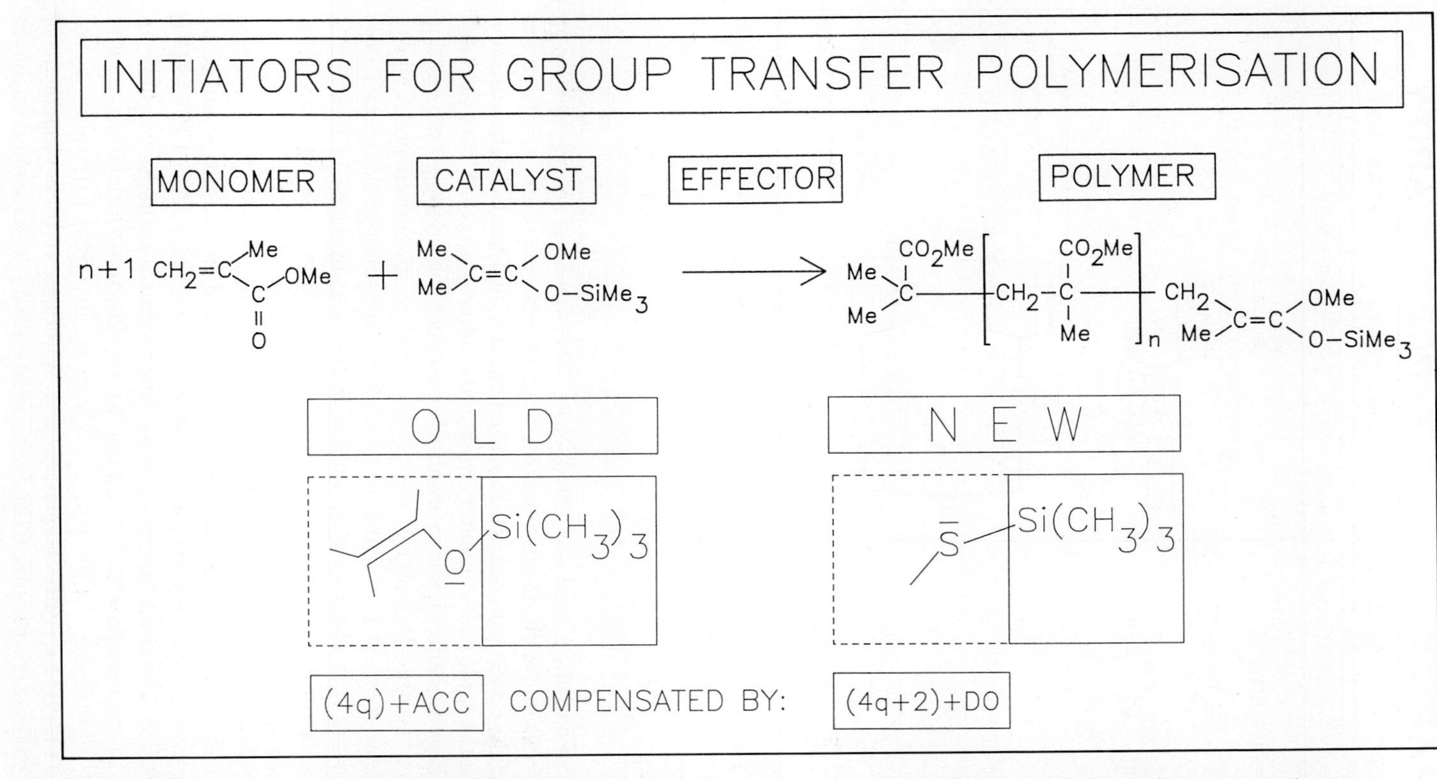

Fig. 8.8. A very simple example of a compensation in a subsystem of two group-transfer effectors shall demonstrate, how to circumvent patents

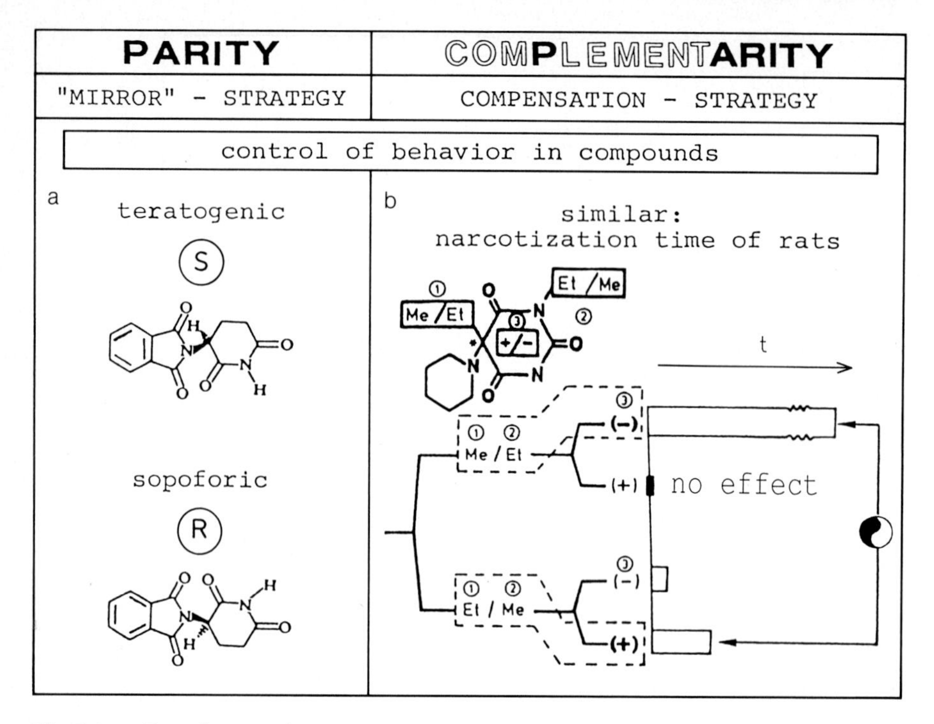

Fig. 8.9. a Enantiomeres have e.g. very different properties as known from the tragedy with thalidomide. (**b**) A change to the alternative stereogenic center can be compensated by alternatives in peripheric alkylgroups

derivative of barbituric acid was tested for its effectiveness as an anaesthetice on rats. As a criterion for the effectiveness the time the rats were of anaesthetized was taken. The (−)-enantiomer exerted a strong effect with a methyl substitution in one and, at the same time, an ethyl group in the other position. A parity change at the stereogenic center leads to absolutely no effect with identical substitution of the parent system. A change of the methyl/ethylgroups in these positions can be compensated by a change in the configuration of the stereogenic center. It only remains to say that alternative variations in the – considered in isolation – "achiral" subsystems invert the antagonistic effect of the enantiomeric subsystems in such a way, that comparable effects are observed" [154].

These experiments of Knabe [235] exert a special fascination. He was able to show, that the effect of one enantiomer is simulated by its mirror image by peripheric variations (corresponding to complementary ALTERNATIVE PRINCIPLES). The consequence of those experiments is perhaps, to assume an inversion of phase-relations – leading to stabilization/destabilization and activation/inhibition – in the whole enzymes by marginal variations of ALTERNATIVE PRINCIPLES at small regions. Without severe rearrangements in the whole enzyme it is possible to understand so-called allosteric phenomena. In the space filling proteins – especially in the membranes of the cell, which are all constructed with building stones with stereogenic centers of one parity – perhaps peripheric variations of the ALTERNATIVE PRINCIPLES of the peripheral subsystems or in areas near the receptor could lead to "allosteric" (perhaps better allophasic) phenomena of the whole system (by changing phase relations by alternative conformations, association/deassociation, protonation/deprotonation, *cis-/trans*-configurations and so on in these peripheral subsystems). So the opening of channels for ions e.g. could be achieved without dramatic reorganization of the proteins of cell membranes being necessary (see Fig. 9.4 and context).

Finally a totally abstract, possible application of compensation strategy should be discussed for the development of new herbicides, drugs and so on [236]. It has been demonstrated that the CONCEPT OF ALTERNATIVE PRINCIPLES can be applied in such a way, that even the thinking in classes of functional groups becomes partially obsolete. We will demonstrate this procedure only in principle and for this task we have chosen some derivatives of the natural herbicides pyrethrine I and II, still on the market, in Fig. 8.10. Again four examples are selected with increasing complexity and abstractness.

Insignificant changes in activity are observed ((a)), when a vinylic subsystem is replaced by a phenylgroup (replacement of one 4q + 2- by a similar 4q + 2-subsystem). A compensation is not necessary and not performed. In (b) two methyl groups are replaced by two ethyl groups at the terminal positions of the double bond, what is compensated by alternative structures (OPEN/CYCLIC). In (c) a carboxylic acid ester group is replaced by a thioester group and this is balanced (by replacement of an ethyl- by a methylgroup and) by "cyclization" (OPEN/CYCLIC: ester/thiolactone) (for further differences see (d)).

In (d) the relatively abstract compensation by keto-/enole-form respectively alternative positions (2+4/1+3-coupling) should be discussed in a little more detail. Just as the "direct unifying" has to be considered to be only one procedure in a thinking model, now the alternative – called "direct splitting" – is applied, too.

By splitting the five-membered ring in 2,4-positions of the oxo-subsystem the corresponding enole-subsystem is then consecutively unified in the alternative 1,3-positions (see for this Figs. 3.20 and 21 as well as Fig. 5.5, too): compensation of aldehyde-/enole-form (4q+2-/4q-) by 2,4-/1,3-position (of a subsystem perturbated in 1-position: see Scheme 3.1). The relation of Me/Et variation in (b) and Me/H replacement in (d) to one of the compensating phenomena is done arbitrarily, but nevertheless based on experience.

In Fig. 8.11a an example is depicted [237] in which, in a relatively complex system an alternative structure (OPEN/CYCLIC) leads to a reverse in the direction of the optical induction in a biomimetic hydrogenation step. In this connection, investigations are of special interest in which the sequence of the messenger, a long-chained protein directly interacting with the receptor, is well-known. Studying the publications [238–240] dealing with the CYCLIC structure of a small protein with the necessary sub-sequence to fit well at the

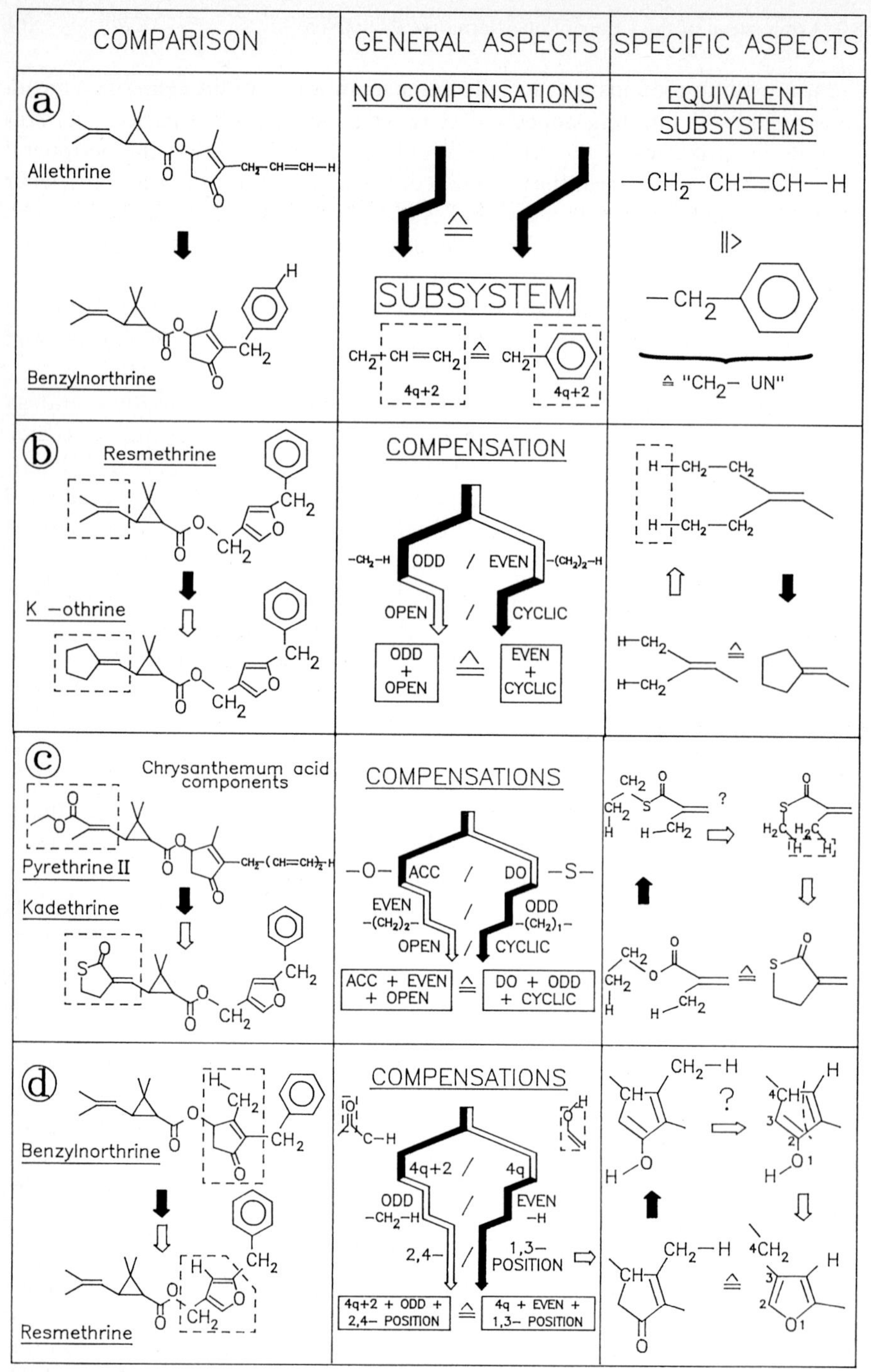

Fig. 8.10. The natural herbicides pyrethrine I and II can be modified in such a way, that in (**a**) no significant change in effectiveness is observed. Other, similar herbicides, which are on the market, are investigated by the CONCEPT OF ALTERNATIVE PRINCIPLES for present compensations. Inversely this procedure offers a new strategy for the synthesis of new pharmaceuticals, drugs and other chemical specialities

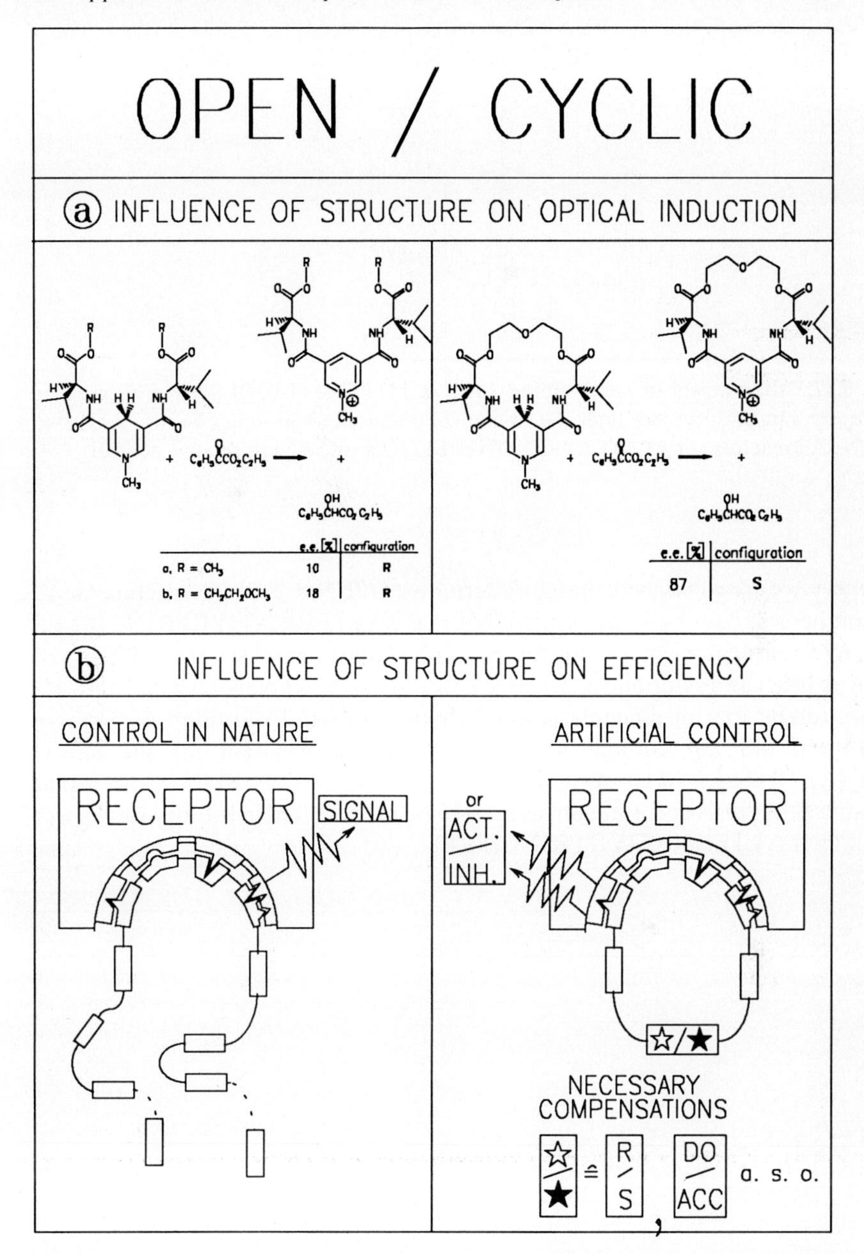

Fig. 8.11. **(a)** The decision OPEN/CYCLIC realized in structures causes, even in complex systems, an alternative sign in the rotational value of the enantiomeric excess. **(b)** Therefore the CONCEPT OF ALTERNATIVE PRINCIPLES should be important too in OPEN/CYCLIC structures of proteins for designed compensations

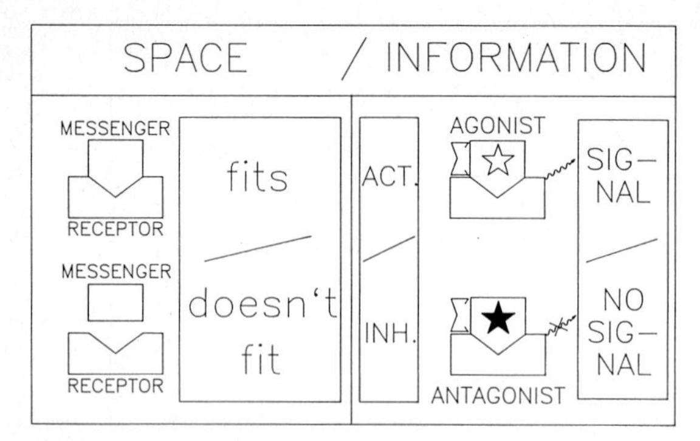

Fig. 8.12. Alternatives in space filling such as IT FITS/IT DOESN'T FIT may be *necessary*, but they are *not sufficient* to control phenomena in time like SIGNAL/NO SIGNAL (absolute) or ACTIVATION/INHIBITION of a signal (relative control)

receptor, we are convinced, that the alternative OPEN/CYCLIC structure has to be compensated for by at least one change in the ALTERNATIVE PRINCIPLES (e.g. *R/S*) of the non-interacting amino-acids. This is demonstrated in Fig. 8.11b in an abstract and symbolic manner. We would like to emphasize, with Fig. 8.12, very strongly that messengers tailored by drug design [241] fulfill a *necessary* condition, but *not* enough to gain *sufficient* information for the effects SIGNAL/NO SIGNAL (absolute) or ACTIVATION/INHIBITION of a signal (relative). Agonists and antagonists have to be investigated in future with respect to all their ALTERNATIVE PRINCIPLES.

9 Information from Alternatives in Biochemistry

"Symmetry as wide or as narrow as you may define its meaning, is one idea by which man through the ages has tried to comprehend and create order, beauty and perfection."

Herman Weyl [17]

The relationship in Fig. 9.1 demonstrate, that it is extremely useful for a preparatively working chemist to investigate the digital decisions of highly organized biosystems, which are determined by chemical systems or by a whole pattern of signals, with respect to the ALTERNATIVE PRINCIPLES in subsystems, especially those with absolute effects. These alternatives, applied in the evolution of simple, chemical systems e.g. striving at the target of catalyzed organic synthesis of even complex compounds with high selectivity, lead to a type of biomimetic strategy. On the other side it is interesting for the biochemist, to identify the differentiations and compensations in the order decisions of simple systems with normally only relative effects.

9.1 Alternative Information in Nucleic Acids and α-Amino-Carboxylic Acids

An interesting comprehensive investigation [242] accounts for the fact, from what variations the differentiating information is derived in the nucleic acids of the genes. In Fig. 9.2 only the aromatic tautomeres of the nucleic acids are investigated, because keto-/enole- like imine-/enamine-forms behave as direct alternatives to each other. In Fig. 9.2a, it is depicted, how the three dimensions in space are determined by basic decisions via two dipoles and choosing only one parity demonstrated in the π-system. The fine tuning arises from substituents in selected positions. In (b) the differences between the pyrimidines and the purines are accentuated, while in (c) the differences in the corresponding complementary bases are depicted. Neglecting the subsystem with an 1,3-diaza-allylic moiety (4q-π-system), the following variations alone result (vide Fig. 2.14):

S	-H		
/	/		
AS	$-CH_3$ /	$-NH_2$ /	-OH
	EVEN /	ODD /	EVEN

which are available for the differentiation of the information.

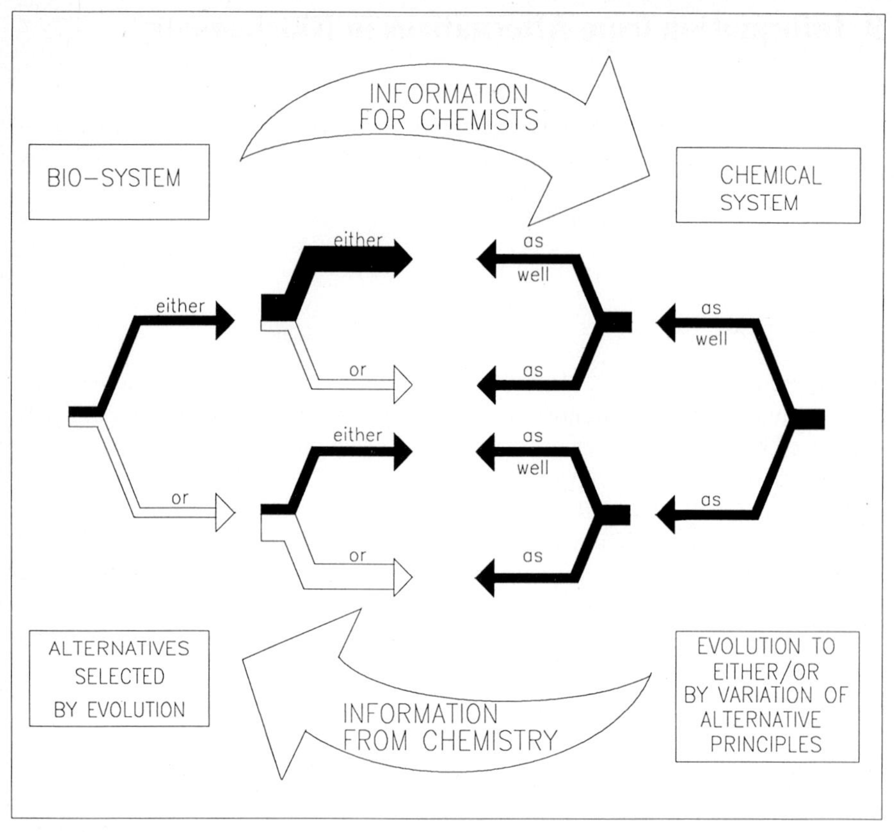

Fig. 9.1. The mutual flow of information between evolved and unevolved chemical systems has attractive aspects

In Fig. 9.3 the three basic decisions to determine the three dimensions for the plane of the amide-subsystem (with six atoms in one plane) [242] are presented. One dimension is controlled by a strong dipole moment (C/O + N), the second dimension by O/N-differentiation (see e.g. Figs. 2.15 to 18) and the third dimension again is characterized by the decision restricting to only one parity. The "joints" of these molecular planes are determined both by the intrinsic ALTERNATIVE PRINCIPLES in the (20) different R-groups of the α-amino-carboxylic acids (Fig. 9.3b) within the chain and by all types of electromagnetic through-space interactions of the subsystems within the chains. The influences of the substituents R in the acids perhaps could be better understood by "dominating" and "slaving" in the cooperativities and compensations of all intrinsic ALTERNATIVE PRINCIPLES present in the R-groups.

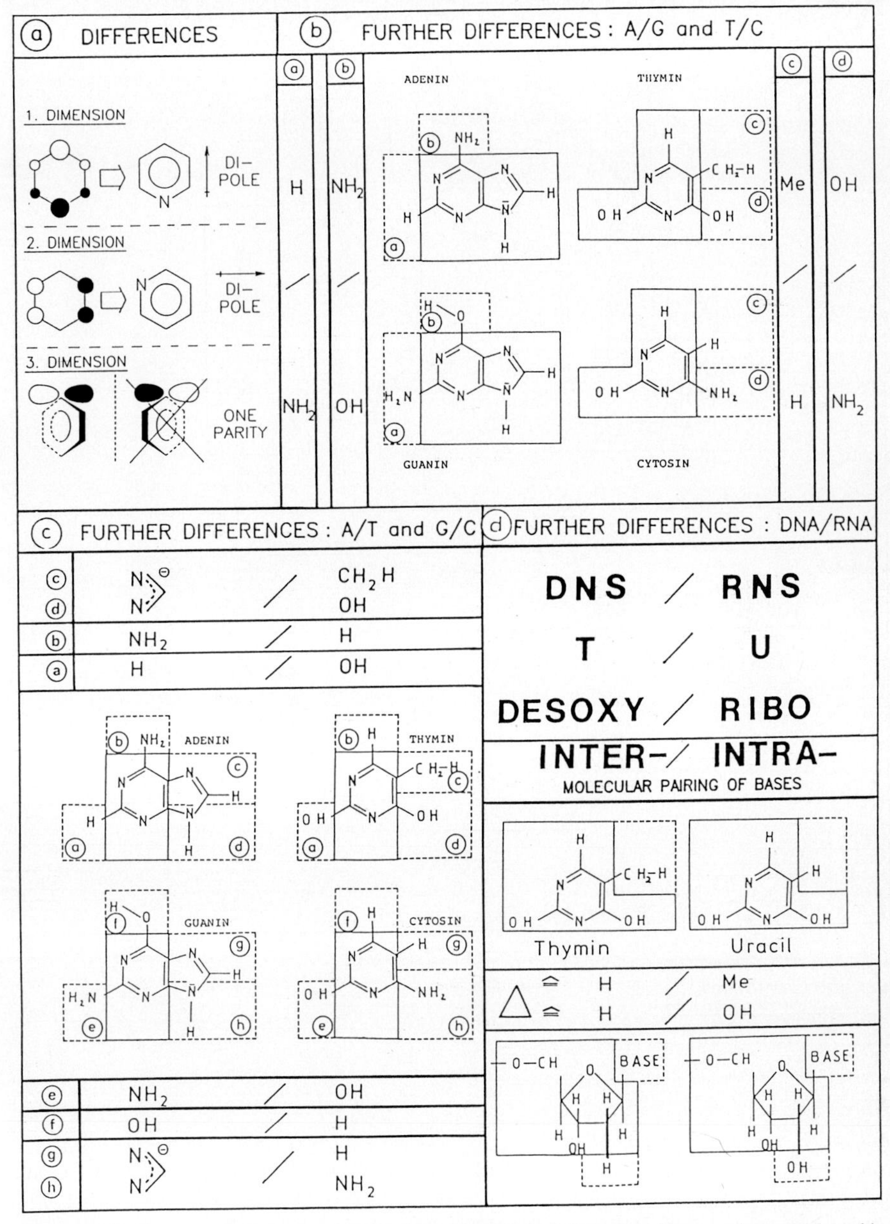

Fig. 9.2a-d. The ALTERNATIVES in the molecules of the genes (here only considered in the tautomeric forms of arenes) are indicated for the three dimensions of space in the system in **(a)**, the additional differences by perturbation *at* the systems caused by changing substituents between the two different pairs in **(b)** and the complementary ones in **(c)** as well as for intra- and intermolecular pairing of bases in **(d)** [242]

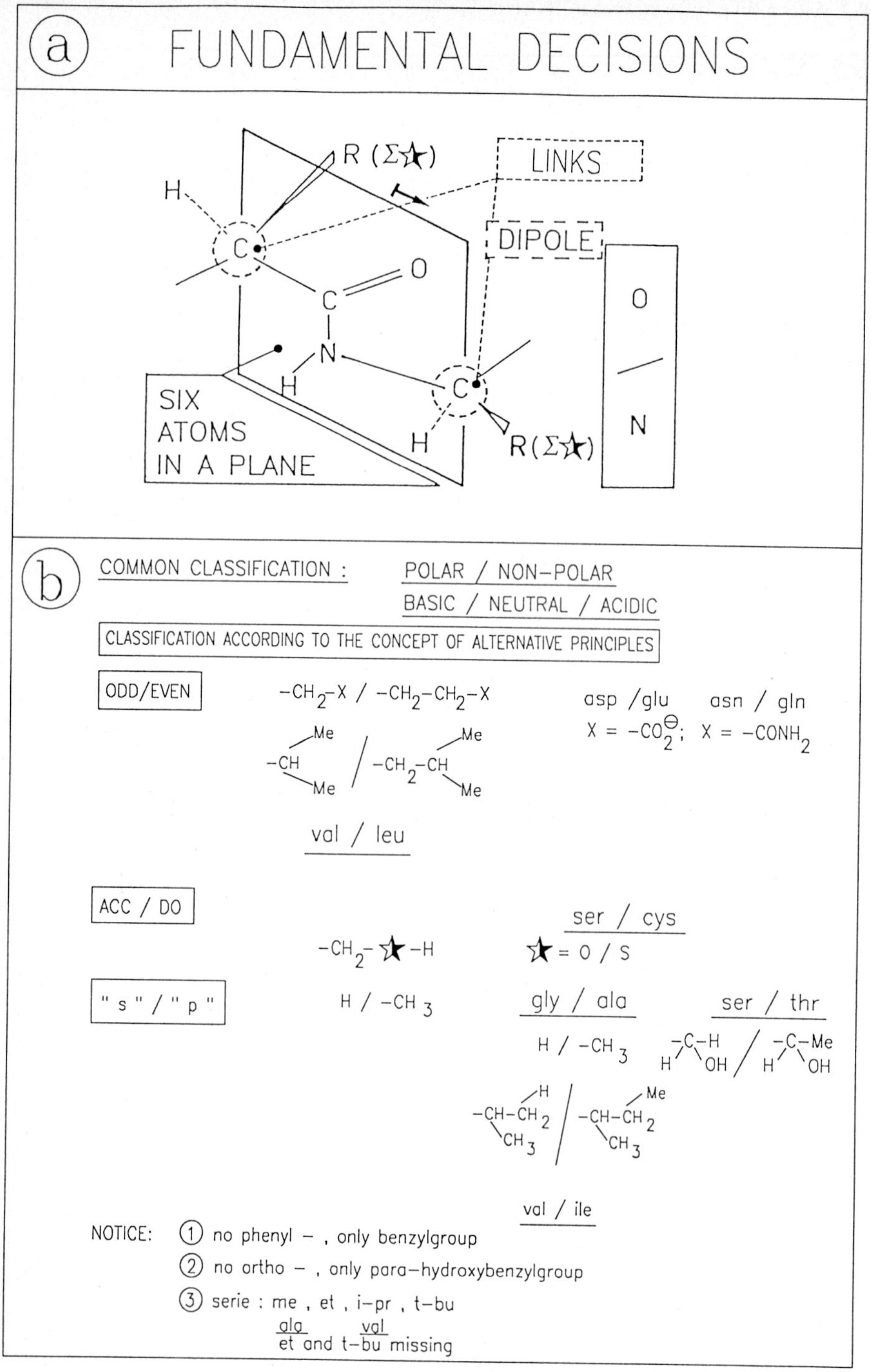

(a) FUNDAMENTAL DECISIONS

R (Σ☆)
LINKS
H
C
DIPOLE
O
C
O
N
H
C
SIX ATOMS IN A PLANE
H
R(Σ☆)
N

(b) COMMON CLASSIFICATION : POLAR / NON-POLAR
BASIC / NEUTRAL / ACIDIC

CLASSIFICATION ACCORDING TO THE CONCEPT OF ALTERNATIVE PRINCIPLES

ODD/EVEN

$-CH_2-X$ / $-CH_2-CH_2-X$

asp /glu asn / gln

$X = -CO_2^{\ominus}$; $X = -CONH_2$

$-CH(Me)_2$ / $-CH_2-CH(Me)_2$

val / leu

ACC / DO

ser / cys

$-CH_2-$☆$-H$

☆ = O / S

" s " / " p "

H / $-CH_3$

gly / ala

ser / thr

H / $-CH_3$

$-C(H)(H)-OH$ / $-C(H)(Me)-OH$

$-CH(H)-CH_2-CH_3$ / $-CH(Me)-CH_2-CH_3$

val / ile

NOTICE: (1) no phenyl – , only benzylgroup
(2) no ortho – , only para–hydroxybenzylgroup
(3) serie : me , et , i–pr , t–bu
ala val
et and t–bu missing

Fig. 9.3a,b. Beside the basic decisions made for space in the building stones of proteins **(a)** the individual ALTERNATIVE PRINCIPLES in the substituents of α-amino-carboxylic acids are considered in **(b)** [242]

9.2 Alternative Information at Bio-membranes

As we tried to show in Fig. 9.4a, some asymmetries in biomembranes exist in alternatives, which are proposed by the CONCEPT OF ALTERNATIVE PRINCIPLES [242]. The biomembranes may be imagined as two-dimensional liquid crystals. Beside the double layer from enantiomerically pure glycerol derivatives with normally two long hydrocarbon chains and a small functionalized group with multifold variations, the membrane exists as a big arsenal of different proteins [243].
Five aspects are selected arbitrarily in Fig. 9.4b which seem important with respect to the CONCEPT OF ALTERNATIVE PRINCIPLES [242]:

1. The two layers are formed by building stones of one parity. The direct consequence is, that the "double layer" can't be build up by enantiomeric layers. The whole "double layer" belongs coherently to only one parity and its molecular analogon could be assumed to be the corresponding enantiomere of tartaric acid as indicated in Fig. 9.5b. The prevailing uniform information in the double layer is demonstrated by the permanently coiled α-helix of membrane proteins (see Fig. 9.5c). In one of the two layers – seen from inside – the spiral runs from the N- to the O- and the opposite layer from the O- to the N-terminated end (see ad (4) in Fig. 9.4b, too). It is interesting to note, that the formation of a double layer by building stones of one parity leads consequently to an asymmetry of the membrane via the proteins.
2. The two long chains, with only an odd number of methylene groups in the long hydrocarbon chains, differ – seen from the stereogenic center of glycerol – in every case by one methylene marked by shadowing, i.e. 50% of the ends of the chains have alternative phase relations to each other. Therefore perhaps a permanent dipole moment cannot be formed by collecting the interactions of induced dipols between the double layer.
3. In highly specialized membranes we are faced with different subsystems inside or outside (3 × H/3 × Me) [244]. Corresponding alternative distributions inside/outside are observed at different subsystems [244].

   A short time ago the arrangement of whole enzyme collectives together with other prostethic groups were identified (the crystalline reaction center of the purpur bacteria Rhodopsendomonas veridis) in their relationship and their cooperativities e.g. for the capture of photons and transformation of the accepted energy by transport of electrons and protons into membrane potential [245]. Thinking about the principle completed process of parity breaking, these systems have of course a almost perfect illusion of symmetry by X-ray analysis, but the fact, that pathways of only one parity (of two series of pigments one is considered as a rudiment of evolution) are open for electrons and protons, should perhaps no longer be surprising.
4. By determining the position of N- and O-atoms in this collective of enzymes it was stated, that inside/outside at the surface area of this center practically only one type of heteroatom (N/O) is present [246].

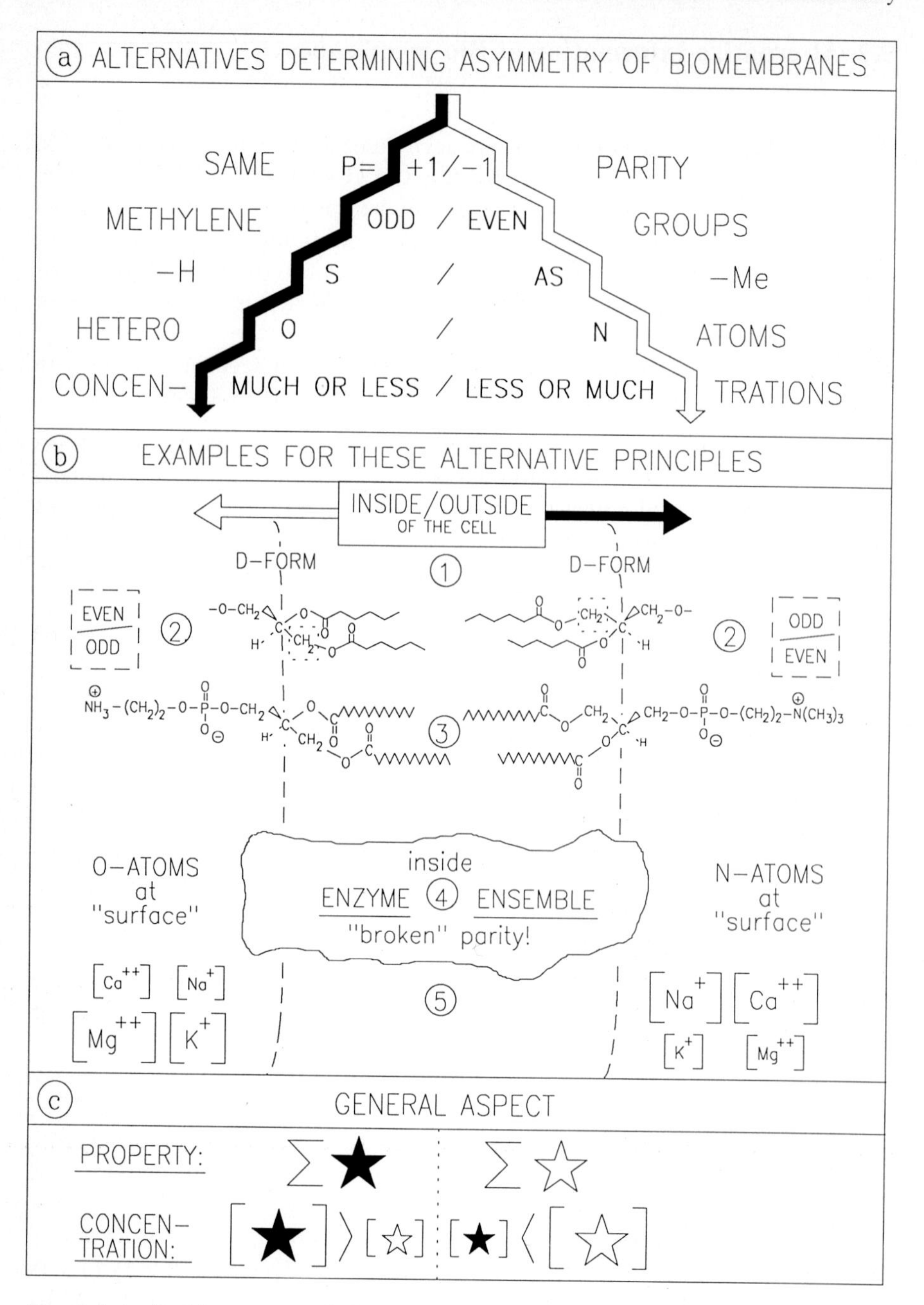

Fig. 9.4a-c. Besides a general decision tree for ALTERNATIVE PRINCIPLES in membranes with building stones of one parity **(a)** some individual real examples **(b)** and a generalization **(c)** are described [242]

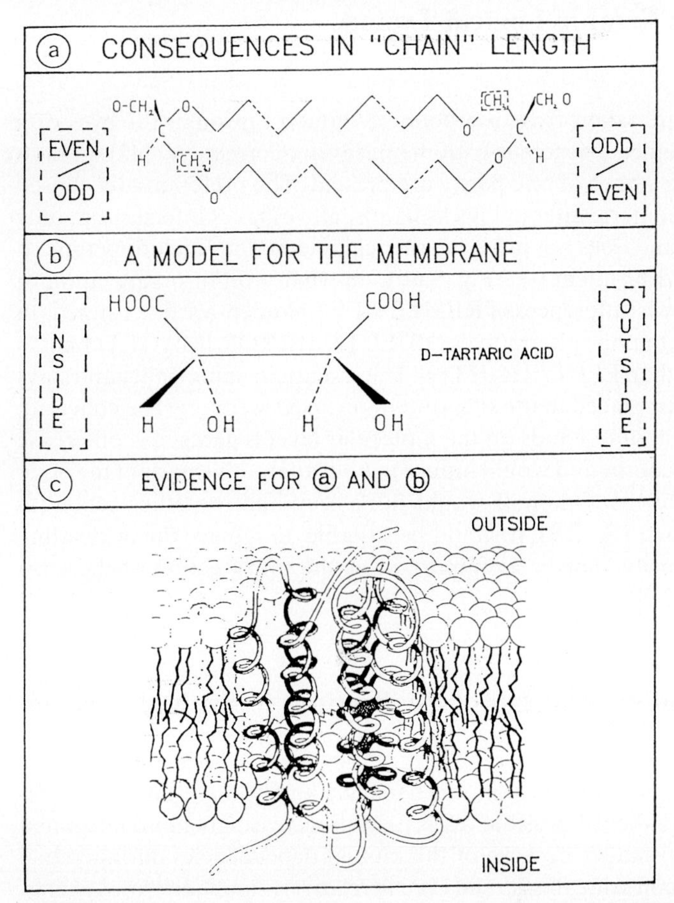

Fig. 9.5a. Beside the hint to the homology in both chains of paritetic building stones of membranes is referred in **(b)** to a good analogous molecular model and in **(c)** to the non-mirror-character of both layers in membranes [242] (see text, too)

5. A further consequence of this asymmetry is, that even under homoeostatic conditions inside and outside the ion concentrations of corresponding ion pairs with agonistic/antagonistic activity like Na^+/K^+ or Ca^{++}/Mg^{++} in every case are alternatively high/low or low/high. The differences of the potentials can be increased dramatically by active pumping [242].

Inside/outside arrangements of enantiomeric compounds means, that one dimension e.g. related to the stereogenic center is inverted. The complementary ALTERNATIVE PRINCIPLES of properties and amounts (vide Fig. 9.4c) may be identified inside/outside – at least relatively. The biomembranes must be considered in a certain sense as order condensators, because they are formed by building stones of one parity exclusively.

9.3 Chirality, an Error in Logical Typing

In Fig. 9.6a a picture taken from the book "Symmetry through the eyes of a chemist" [247]. is depicted together with the necessary correction [242]. In both hands only building stones of one parity are present. The preferentially chosen analogy for diastereomeric interactions as hands/gloves takes into account only the aspect of space and does not justice to the real interactions including time. It is still an unresolved problem (see Fig. 9.6b), on what level of organization of matter the well-known differences of RIGHT/LEFT handers are determined. In (c) it is accentuated, that the alternatives OBJECT/MIRROR IMAGE belong to a different category than LEFT/RIGHT [9]. The essentially same right hand stays on writing in a mirror placed to the side (but in reversed-writing). The coherent, unequivocal parity in both hands on the molecular level is necessary, otherwise every enantiomeric compound would induce in both alternative parts of the body diastereomeric problems. One half would fall asleep and the other gets teratogenic problems! (see Fig. 8.9). It would be suitable, to replace the misleading term chirality by parity. This is as important as the introduction of the term stereogenic center [248].

9.4 Restriction of the Number of Realizations in Evolved Systems

Riedl [8] has pointed in a popular scientific book to a highly interesting phenomenon. Thereby the number of possible realization in increasingly more integrated biosystems becomes smaller because of the mutual dependencies on individual alternatives. In a certain sense this is valid even in relatively simple enzyme systems, of which the efficiency is describable by a few parameters in a distinct manner [169]. A proof of this observation is also demonstrated by order parameters, which, according to Haken [232] cast light on the self-organization of highly complex systems in their cooperativities with feed back.

We have restricted ourselves quite pragmatically to ALTERNATIVE PRINCIPLES, which should allow a qualitative strategy for the evolution of chemical systems on the level of organic synthesis. We discuss the possible multi-dual decisions, starting from double-dual controls (for avoiding "problems of three" see e.g. Fig. 8.3). On the other hand, A. Heimbach has pointed to the fact [242] (see Scheme 9.1), that Mendel's second law could be interpreted as an idealized law for alternative cooperativities (WHITE or RED) and compensations (2 x ROSE) of classes of ALTERNATIVE PRINCIPLES represented in the genes. The multidual bifurcations are reduced by evolution to three results (1:2:1).

Much easier equations than the last one could be solved by applying the CONCEPT OF ALTERNATIVE PRINCIPLES. The selection of a certain α-amino-carboxylic acid is determined by a codon of nucleic acids for the next

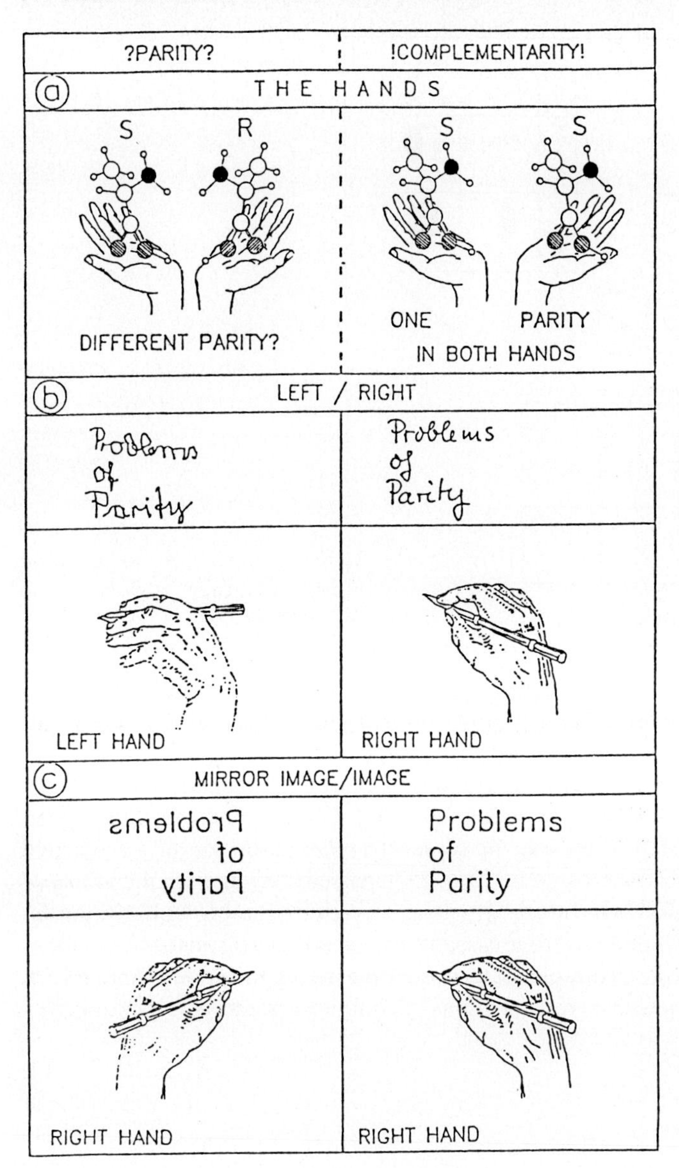

Fig. 9.6a-c. The following errors in logical typing are presented. **(a)** The hands are not enantiomeric, but complementary. **(b)** The left and right hand differ qualitatively and quantitatively. **(c)** The right hand is still writing in the mirror (but reversed writing) [242]

step in the synthesis of proteins. But for some α-amino-carboxylic acids there are multifold codons with differences in the necesary three nucleic acids ("degeneration"). Here it would be interesting to study by which ALTERNATIVE PRINCIPLES at all the "degenerated" codons differ perhaps with the intention, in what different states the α-amino-carboxylic acids are eventually

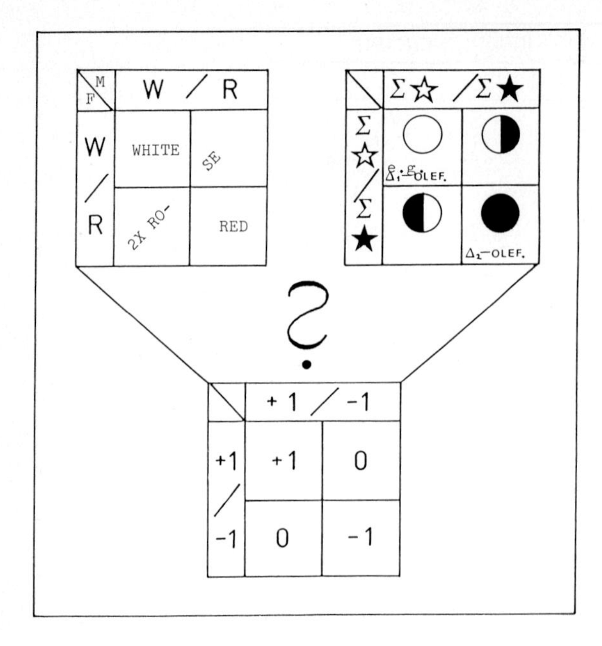

Scheme 9.1

incorporated into the proteins. It might be possible, that "Nature" does not waste information.

But an essential limitation for realizations is often observed in just such systems, which are strongly determined by preorganization. The efforts, to simulate e.g. an individual reaction step of biosystems by severe preorganization of a restrictive spatial arrangement in macrosystems [249, 250], passed a zenith with the award of the Nobel prizes in 1987. In future, it will be more important to understand the selforganization of whole alternative cascades of reaction steps and to control them systematically. These last examples of too restrictive preorganization are not meant here, whenever the decrease in realizations of complex systems was mentioned.

10 Acknowledgements and Petition

Jeder Naturwissenschaftler, der diesen Namen verdient, unterstellt sich aber der Kontrolle durch das Experiment oder durch den Formalismus einer logisch konsistenten und empirisch bewährten Theorie

H. Primas u. U. Müller-Herold [227]

Even an experimenter will not start a systematic investigation e.g. of a reaction without the "benefit of prejudice" [251]. Our experiences with catalytic systems were directed by experiments. This and our efforts, not to forget essential, fundamental aspects of theoretical chemistry and to apply them, guided us to the pragmatic, qualitative CONCEPT OF ALTERNATIVE PRINCIPLES. The ambitions aim of this book was, if possible, to avoid errors in logical typing. It was hardly to adhere to, because "experimenter often combine empirical regularities and deeper theoretical insights in an artistic manner. Any such representation of models relies heavily on not-formalized and unconsidered, pre-understanding, therefore it "– the phenomenological conception of models –" should be used only in combination with 'common sense' " [227].

Should this common sense have played a nasty trick on us, we have to beg that the theorists draw our attention to this. Projected into the next years this concept of shape forming caused by alternatives should prove itself to be a possibly new pathway to general chemistry – without burdering those who take an interest in chemistry in its widest sense, with too high mathematical demands of theoretical chemistry or even to overstrain. In addition general, epistemological problems should be included for the qualitative understanding of complex chemical systems. Thereby the alternative experiment should be placed in the foreground of interest. This prevents one-sided dogmatism.

Developing the concept slowly we consciously followed strategies proposed by Corey [252] and summarized in Scheme 10.1 at the end of the chapter. In 1960 one of us (P.H.), received a first important hint from K. Ziegler when he corrected the thesis (vide Fig. 10.1 above) by a comment on parities in subsystems, where the corresponding whole systems are different in enthalpy of formation. A long time passed before the aha-experience, which was finally induced by the digital descriptions of parities and local parities by Ugi and coworkers [228], schematically presented in Fig. 10.1 below (for consequences, see Sect. 5.3).

K. Wisseroth [40,253,254] of BASF (vide Fig. 10.2a) presented us with a deep insight into the basic phenomenon of catalysis to be seen as a time-dependent resonance by means of analogous simple models of increasing complexity (sympathetic pendulum, coupled electric circuits, two coupled Schrödinger equations and so on). The schematic presentation in Fig. 10.2b should induce the reader to study the considerations concerning efficiency of enzymatic systems

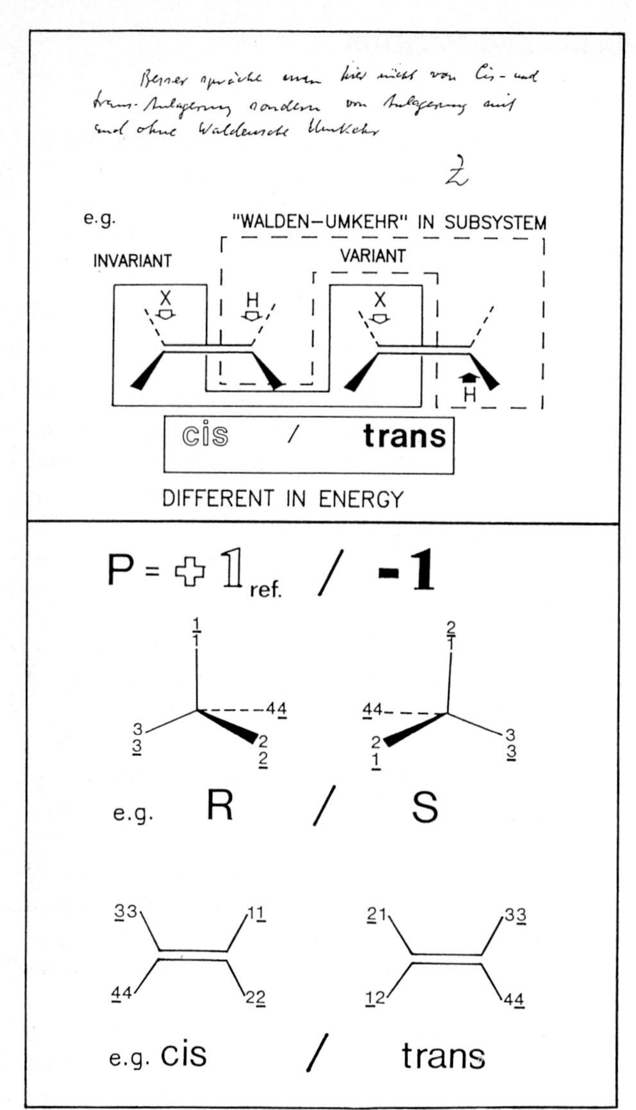

Fig. 10.1. Above – Hint given by K. Ziegler (1960, when correcting the thesis of P.H.) to inverse parity in subsystems of an olefine. Below – Digital representation for problems of stereochemical alternatives with the parity P = +1/−1 [228]

with the convincing model of how to develop a "unified bonding" [169]. In (c) we pointed to the excellent ideas of Haken [1] in order to activate the creativity concerning self-organization of matter, which encouraged us to try with the help of the qualitative CONCEPT OF ALTERNATIVE PRINCIPLES to evolve the establishing alternative orders in chemical systems "from the top to the bottom" [255] – as to be seen in an open system – from differentiating and compensating alternatives.

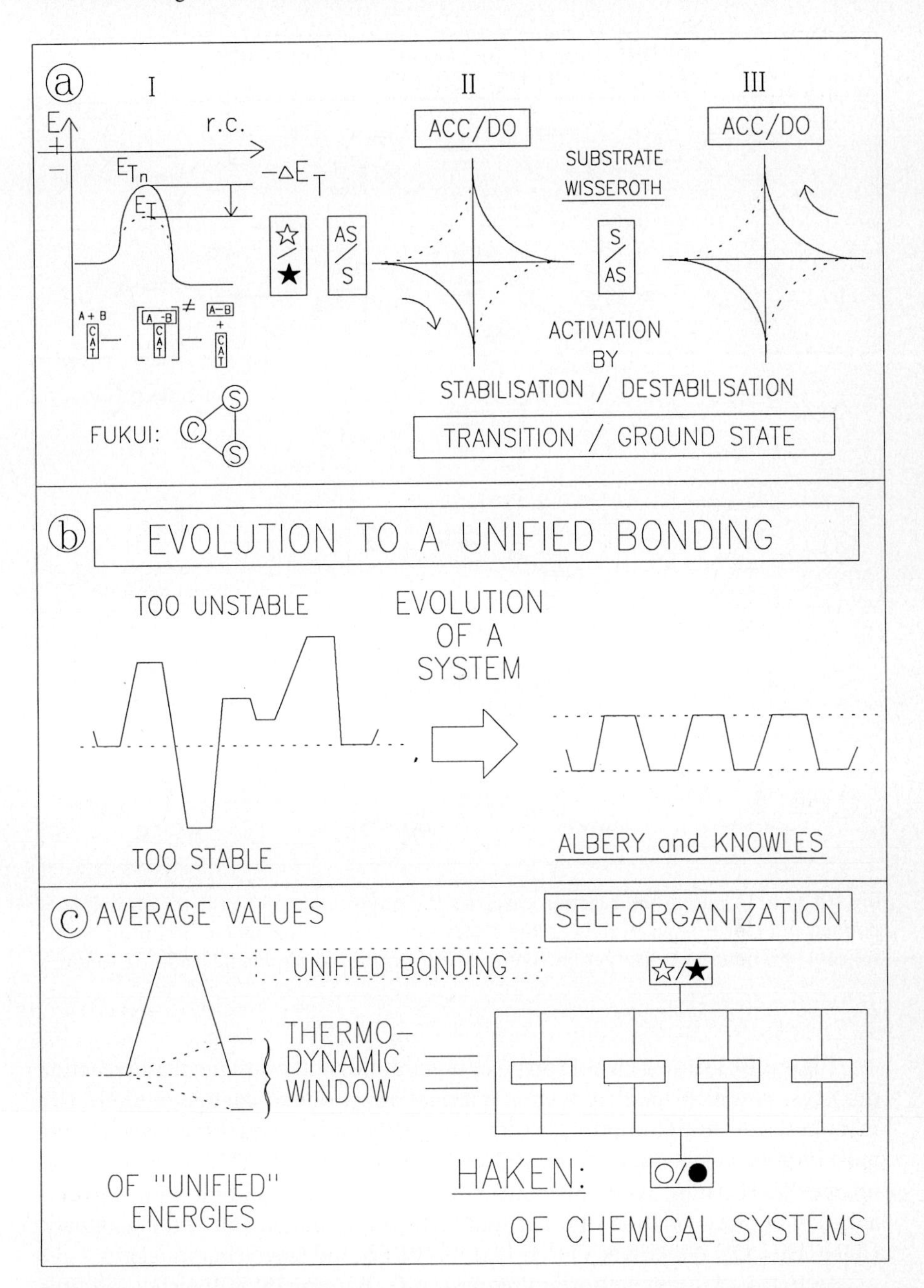

Fig. 10.2a. Hint given by Wisseroth to the mutual coupling of DO/ACC and S/AS in a model for catalysis. **(b)** The concept of unified bonding by Albery and Knowles points to a coupling of thermodynamics and kinetics. **(c)** The individual modes (reactions) are controlled by order parameters referring to Haken's concept of synergetics

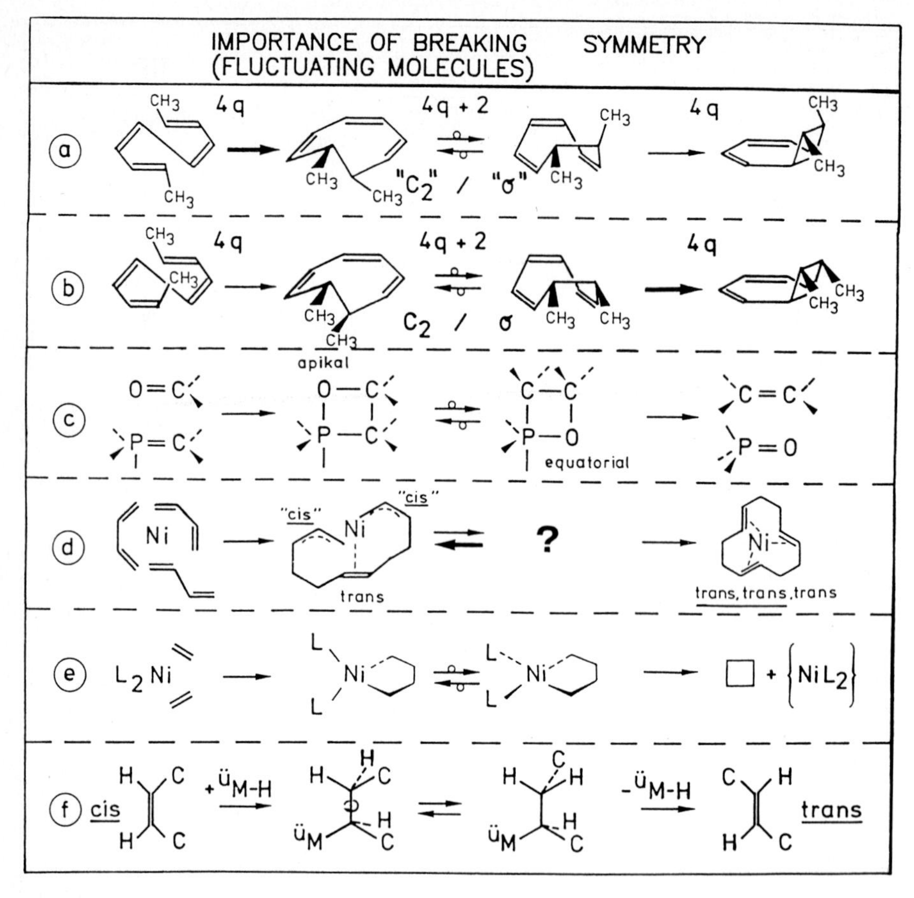

Fig. 10.3a-f. In individual reaction cascades the importance of symmetry inversions is pointed out (for this see Halevi [226]). Every individual is considered according to the concept of symmetry conservation. Symmetry inversion is omitted and leads to irritations

Figure 10.3 shows a whole series of possibilities and prerequisites in reaction cascades, in which small activation energies allow one to change in a selective manner to alternative orders. If we have correctly understood this inversion is the most important statement of the OCAMS-method of Halevi [226] which widens up the W.-H.-rules. And this controlled inversion in addition with a generalization of phase relation rules for small effects is the main aim of the pragmatic CONCEPT OF ALTERNATIVE PRINCIPLES, too (see Scheme 7.1).

In Fig. 10.4a, very important request of G. Bateson [9] is depicted – a little bit modified from our point of view. The necessary processes of the understanding by separation of the terms in the levels are opposed to the levels of mathematical formulas and formalizations of Russell and Whitehead [256]. We have chosen an evolving spiral of epistemology as mediator between the levels instead of the

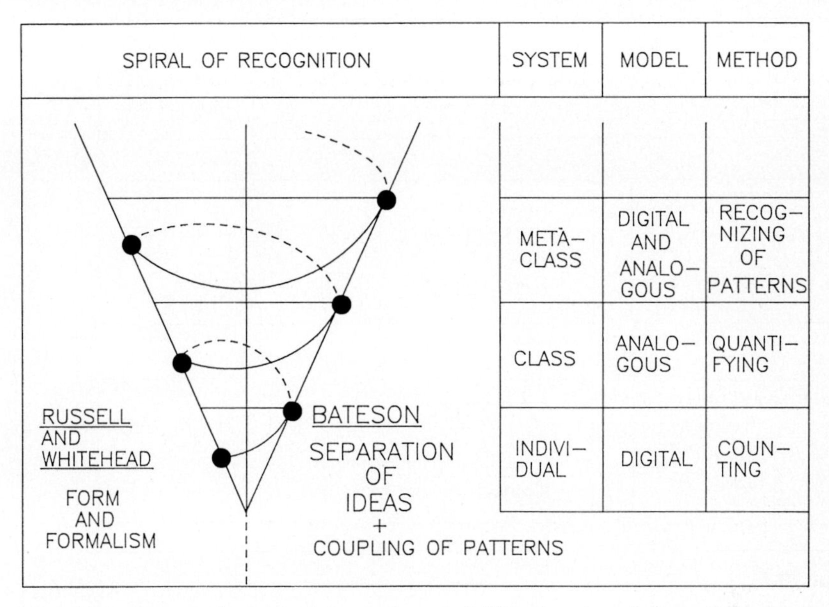

Fig. 10.4. The mathematical formalisms of Whitehead and Russell in the levels of increasing complexity are confronted with the processes of understanding and defining the differences between the levels. We have chosen a spiral of recognizing instead of a didactic ladder proposed by Bateson. The terms for increasing complexity are presented for simple examples of systems, models and methods

dialectic alternating ladder proposed by Bateson. Some terms for the individual levels concerning systems, models and methods are placed together and were applied, hopefully without errors in logical typing. Describing general chemistry one is faced in addition with the problem of selecting from an ocean of facts only those examples, which – "as a pattern that connects" [9] – offer as many connecting aspects as possible.

The temperature-dependent Bouduard equilibrium, for example should never be discussed without the three important technical processes: at low (400°C) and high temperature (1100°C). It is surely worth mentioning, that quite similar slags are formed in the blast furnace during Fe-production and in phosphorus synthesis, once by adding lime (to bind the silicates) or sand (to bind Ca).

The chemistry as a mediator between physics and biosciences should not give up its independence by overestimating the value of the mathematics of theoretical chemistry – especially in high school education and even in undergraduate studies. The handicraft of changing substances and materials coupled with the inspiration of interpreting observations – always controlled by the next systematic experiment – must remain a postulate in education by chemistry. This gives a satisfactory account of chemistry as an empirical, experimental science of

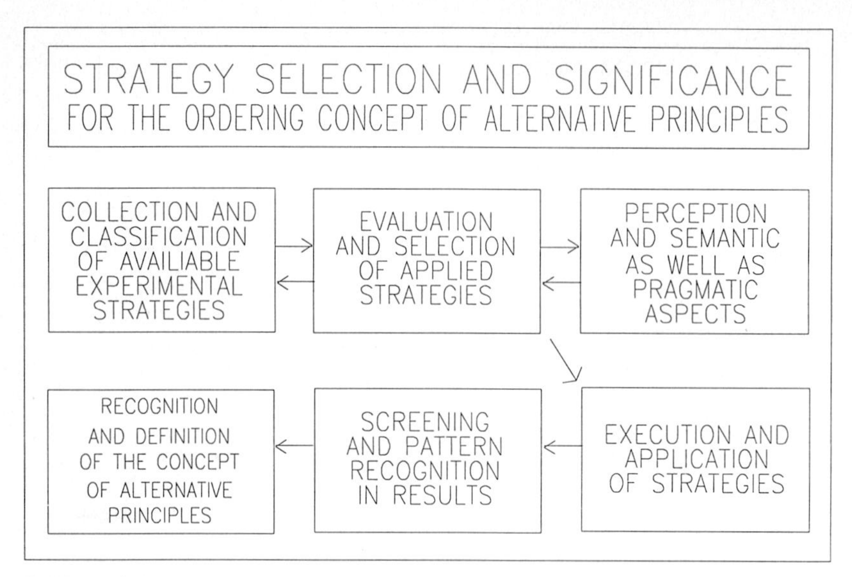

Scheme 10.1

systematic handling of complex systems. To present chemistry only as a further application of mathematics has no chance with respect to physics. All chemists, not only those in theoretical chemistry, must carefully avoid errors in logical typing and one-sided dogma despite the complexity, because, as remarked by Primas [228], "that those who are the most intelligent are just the ones who immediately recognize the logical inconsistency and intellectual dishonesty of such considerations and turn their backs on the subject".

The success of – in principle redundant – terms should not lead to the consequence of conserving them, in spite of errors of logical typing because we are familiar with the contradictions.

We, ourselves, have always respected "control by the experiment" [228], but we are not sure, that we considered all formalisms [257] and prover theories [258].

11 Appendix

Scheme 11.1

			σ_I	P	Δ
ACC / DO	≙	-OCH_3	+0,30	–3,61	–0,80
		-SCH_3	+0,30	–3,68	+0,12
ACC / DO	≙	-CMe_3	–0,01	+4,42	+0,32
		-$SiMe_3$	–0,11	+4,15	–0,89

Table 11.1. New Symbols for parameters

Ref. [59]	This book
$^{FT}\chi$	$^{FT}\chi$
$\Delta^{FT}\chi$	P_L
$\Delta^{FT}\chi_i$	P
$\Delta\Delta^{FT}\chi_{PPh_2X}$	Δ_b
$\Delta\Delta^{FT}\chi_{PPhX_2}$	Δ_w
$\Delta\Delta^{FT}\chi_M$	Δ

Table 11.2 Parameter sets for ligands and substituents (n. b. = not determined)

Nr.	X	$^{FT}\chi_i$ cm^{-1}	P_L cm^{-1}	P cm^{-1}	Δ_w cm^{-1}	Δ_b cm^{-1}	Δ cm^{-1}	$\Delta\Delta$ cm^{-1}
25	$SnMe_3$	–0,32	14,20	4,73	–0,65	–0,95	–0,80	0,30
1	CMe_3	0	13,25	4,42	0,53	0,11	0,32	0,42
26	$GeMe_3$	0,05	13,10	4,37	–0,17	–0,52	–0,35	0,35
21	$SiMe_3$	0,27	12,45	4,15	–0,71	–1,06	–0,89	0,35
46	*c*-Hexyl	0,47	11,85	3,95	n.b.	n.b.	n.b.	n.b.
47	*o*-O-Me-Ph	0,57	11,55	3,85	1,00	0,90	0,95	0,10
41	N(*n*-Bu)$_2$	0,77	10,95	3,65	1,70	1,11	1,41	0,59
40	N(*n*-Pr)$_2$	0,70	10,55	3,52	1,62	1,06	1,34	0,56
42	N(*i*-Bu)$_2$	1,02	10,20	3,40	1,85	1,05	1,45	0,80
39	N(Et)$_2$	1,03	10,15	3,38	1,66	1,05	1,36	0,61
2	*i*-Pr	1,15	9,80	3,27	0,78	0,86	0,82	–0,08
28	CH_2-$SiMe_3$	1,20	9,65	3,22	0,91	0,45	0,68	0,46
48	*n*-Dodecyl	1,67	8,25	2,75	n.b.	n.b.	n.b.	n.b.
3	*n*-Hexyl	1,67	8,25	2,75	0,69	0,39	0,54	0,30
49	*n*-Octyl	1,67	8,25	2,75	n.b.	n.b.	n.b.	n.b.
50	*s*-Bu	1,72	8,10	2,70	n.b.	n.b.	n.b.	n.b.
4	*n*-Bu	1,75	8,00	2,67	0,68	0,51	0,60	0,17
51	*p*-NMe_2-Ph	1,75	8,00	2,67	–0,12	–0,19	–0,16	0,07
5	*n*-Pr	1,80	7,85	2,62	0,78	0,51	0,65	0,27
34	*i*-Bu	1,90	7,55	2,52	0,78	0,42	0,60	0,36
38	N(Me)$_2$	1,98	7,30	2,43	1,19	0,99	1,09	0,20
6	Et	2,10	6,95	2,32	0,68	0,36	0,52	0,32
27	CH_2-*t*-Bu	2,38	6,10	2,03	0,30	–0,01	0,15	0,31
7	Me	2,85	4,70	1,57	0,48	0,41	0,44	0,07
45	*c*-Pr	2,95	4,40	1,47	0,48	0,52	0,50	–0,04
8	Benzyl	3,45	2,90	0,97	0,58	0,01	0,30	0,57
9	Allyl	3,50	2,75	0.92	0,18	0,06	0,12	0,12
52	*p*-OMe-PH	3,50	2,75	0,92	–0,12	–0,08	–0,10	–0,04
10	*o*-Tol	3,55	2,60	0,87	0,38	0,36	0,37	0,02
43	N (Benzyl)$_2$	3,63	2,35	0,78	0,87	0,68	0,78	0,19
53	*p*-*t*Bu-Ph	3,65	2,30	0,77	0.03	–0,13	–0,05	0,16
54	Morpholino	3,73	2,05	0,68	n.b.	n.b.	n.b.	n.b.
55	1-Naphtyl	3,93	1,45	0,48	n.b.	n.b.	n.b.	n.b.
56	*p*-SMe-Ph	4,23	0,55	0,18	–0,13	–0,12	–0,13	–0,01
31	O-*t*-Bu	4,32	0,30	0,10	0,12	0,69	0,41	–0,57
11	Ph	4,42	0	0	0	0	0	0
29	CH_2-S-Me	4,95	–1,60	–0.53	–0,08	–0,10	–0,09	0,02
30	CH_2-O-Me	5,20	–2,35	–0,78	–0,07	0,07	0	–0,14
57	*p*-F-Ph	5,23	–2,45	–0,82	n.b.	n.b.	n.b.	n.b.
58	O-Neomenth	5,37	–2,85	–0,95	0,59	0,55	0,57	0,04
44	N(*i*-Pr)$_2$	5,47	–3,15	–1,05	6,67	–3,67	1,50	10,34
59	*p*-Cl-Ph	5,60	–3,55	–1,18	n.b.	n.b.	n.b.	n.b.
60	O-Menth	5,63	–3,65	–1,22	–0,09	0,13	0,02	–0,22
61	O-Bornyl	5,77	–4,05	–1,35	n.b.	n.b.	n.b.	n.b.

Table 11.2 *(continued)*

Nr.	X	$^{FT}\chi_i$ cm^{-1}	P_L cm^{-1}	P cm^{-1}	Δ_w cm^{-1}	Δ_b cm^{-1}	Δ cm^{-1}	$\Delta\Delta$ cm^{-1}
12	O-*c*-Hexyl	6,00	-4,75	-1,58	-0,57	-0,19	-0,38	-0,38
62	*m*-Cl-Ph	6,13	-5,15	-1,72	n.b.	n.b.	n.b.	n.b.
13	O-*c*-Pentyl	6,17	-5,25	-1,75	-0,51	-0,16	-0,34	-0,35
32	O-*i*-Pr	6,37	-5,85	-1,95	-0,44	-0,15	-0,30	-0,29
35	O-CH_2tBu	6,80	-7,15	-2,38	n.b.	n.b.	n.b.	n.b.
63	O-2Me-*n*-Bu	6,82	-7,20	-2,40	-0,71	-0,25	-0,48	-0,46
14	O-*n*-$C_{18}H_{37}$	6,88	-7,40	-2,47	-0,73	-0,52	-0,63	-0,21
36	O-*i*-Bu	6,90	-7,45	-2,48	n.b.	n.b.	n.b.	n.b.
15	O-*n*-Bu	6,95	-7,60	-2,53	-0,72	-0,44	-0,58	-0,28
16	O-*n*-Pr	6,97	-7,65	-2,55	-0,66	-0,36	-0,51	-0,30
17	O-Et	7,20	-8,35	-2,78	-0,72	-0,44	-0,58	-0,28
64	Et-CN	7,45	-9,10	-3,03	n.b.	n.b.	n.b.	n.b.
33	S-*i*-Pr	7,48	-9,20	-3,07	-0,13	0,24	0,06	-0,37
24	S-Et	7,68	-9,80	-3,27	-0,04	0,21	0,09	-0,25
81	C(0)-*t*-Bu	7,70	-9,85	-3,28	n.b.	n.b.	n.b.	n.b.
37	Se-Ph	7,73	-9,95	-3,32	0,32	0,28	0,30	0,04
18	O-Me	8,03	-10,85	-3,62	-1,03	-0,57	-0,80	-0,46
23	S-Me	8,10	-11,05	-3,68	0,03	0,20	0,12	-0,17
22	S-Ph	8,12	-11,10	-3,70	0,81	0,40	0,61	0,41
65	C(O)-Ph*	8,38	-11,90	-3,97	n.b.	n.b.	n.b.	n.b.
66	O-*o*-Tolyl	9,68	-15,80	-5,27	n.b.	n.b.	n.b.	n.b.
67	*o*-Ph-OPh	9,72	-15,90	-5,30	5,30	n.b.	n.b.	n.b.
19	O-Ph	10,07	-16,95	-5,65	-0,46	-0,04	-0,25	-0,42
68	*o*-*t*-Bu-OPh	10,17	-17,25	-5,75	n.b.	n.b.	n.b.	n.b.
69	*p*-Cl-OPh	11,03	-19,85	-6,62	n.b.	n.b.	n.b.	n.b.
70	*p*-CN-OPh	12,62	-24,60	-8,20	n.b.	n.b.	n.b.	n.b.
20	Cl	16,00	-34,75	-11,58	-0,02	-0,19	-0,11	-0,17

*here we found an unusual broad signal for the symmetric vibration

Table 11.3. Parameter sets of ligands with different substituents (n.b. = not determined)

		PX_3		PX_2Ph		$PXPh_2$	
Nr.	X	$^{FT}\chi$ cm^{-1}	P_L cm^{-1}	$^{FT}\chi$ cm^{-1}	P_L cm^{-1}	$^{FT}\chi$ cm^{-1}	P_L cm^{-1}
25	$SnMe_3$	-0,95	14,20	3,15	10,10	7,60	5,65
1	CMe_3	0	13,25	4,95	8,30	8,95	4,30
26	$GeMe_3$	0,15	13,10	4,35	8,90	8,40	4,85
21	$SiMe_3$	0,80	12,45	4,25	9,00	8,05	5,20
46	*c*-Hexyl	1,40	11,85	n.b.	n.b.	n.b.	n.b.
47	*o*-O-Me-Ph	1,70	11,55	6,55	6,70	10,30	2,95
41	$N(n\text{-}Bu)_2$	2,30	10,95	7,65	6,60	10,70	2,55
40	$N(n\text{-}Pr)_2$	2,70	10,55	7,85	5,40	10,80	2,45
42	$N(i\text{-}Bu)_2$	3,05	10,20	8,30	4,95	10,90	2,35
39	$N(Et)_2$	3,10	10,15	8,15	5,10	10,90	2,35
2	*i*-Pr	3,45	9,80	7,50	5,75	10,85	2,40
28	$CH_2\text{-}SiMe_3$	3,60	9,65	7,70	5,55	10,50	2,75
48	*n*-Dodecyl	5,00	8,25	n.b.	n.b.	n.b.	n.b.
3	*n*-Hexyl	5,00	8,25	8,45	4,80	10,90	2,35
49	*n*-Octyl	5,00	8,25	n.b.	n.b.	n.b.	n.b.
50	*s*-Bu	5,15	8,10	n.b.	n.b.	n.b.	n.b.
4	*n*-Bu	5,25	8,00	8,60	4,65	11,10	2,15
51	$p\text{-}NMe_2\text{-}Ph$	5,25	8,00	7,80	5,45	10,40	2,85
5	*n*-Pr	5,40	7,85	8,80	4,45	11,15	2,10
34	*i*-Bu	5,70	7,55	9,00	4,25	11,15	2,10
38	$N(Me)_2$	5,95	7,30	9,50	3,75	11,80	1,45
6	Et	6,30	6,95	9,30	3,95	11,30	1,95
27	$CH_2\text{-}t\text{-}Bu$	7,15	6,10	9,50	3,75	11,20	2,05
7	Me	8,55	4,70	10,60	2,65	12,10	1,15
45	*c*-Pr	8,85	4,40	10,80	2,45	12,30	0,95
8	Benzyl	10,35	2,90	11,90	1,35	12,30	0,95
9	Allyl	10,50	2,75	11,60	1,65	12,40	0,85
52	*p*-OMe-Ph	10,50	2,75	11,30	1,95	12,25	1,00
10	*o*-Tol	10,65	2,60	11,90	1,35	12,75	0,50
43	$N(Benzyl)_2$	10,90	2,35	12,55	0,70	13,15	0,10
53	*p*-*t*Bu-Ph	10,95	2,30	11,75	1,50	12,35	0,90
54	Morpholino	11,20	2,05	n.b.	n.b.	n.b.	n.b.
55	l-Naphtyl	11,80	1,45	n.b.	n.b.	n.b.	n.b.
56	*p*-SMe-Ph	12,70	0,55	12,75	0,50	12,95	0,30
31	O-*t*-Bu	12,95	0,30	13,15	0,10	13,85	-0,60
11	Ph	13,25	0	—	—	—	—
29	$CH_2\text{-}S\text{-}Me$	14,85	-1,60	14,25	-1,00	13,70	-0,45
30	$CH_2\text{-}O\text{-}Me$	15,60	-2,35	14,75	-1,50	14,10	-0,85
57	*p*-F-Ph	15,70	-2,45	n.b.	n.b.	n.b.	n.b.
44	$N(i\text{-}Pr)_2$	16,40	-3,15	22,05	-8,80	10,65	2,60
59	*p*-Cl-Ph	16,80	-3,55	n.b.	n.b.	n.b.	n.b.
60	O-Menth	16,85	-3,65	15,55	-2,30	14,75	-1,50

Table 11.3. *(continued)*

		PX_3		PX_2Ph		$PXPh_2$	
Nr.	X	$^{FT}\chi$ cm^{-1}	P_L cm^{-1}	$^{FT}\chi$ cm^{-1}	P_L cm^{-1}	$^{FT}\chi$ cm^{-1}	P_L cm^{-1}
61	O-Bornyl	17,30	–4,05	n.b.	n.b.	n.b.	n.b.
12	O-*c*-Hexyl	18,00	–4,75	15,85	-2,60	14,65	-1,40
62	*m*-Cl-Ph	18,40	-5,15	n.b.	n.b.	n.b.	n.b.
13	O-*c*-Pentyl	18,50	-5,25	16,25	-3,00	14,85	-1,60
32	O-*i*-Pr	19,05	-5,85	16,70	-3,45	15,05	-1,80
35	O-CH_2-*t*Bu	20,40	-7,15	n.b.	n.b.	n.b.	n.b.
14	O-*n*-$C_{18}H_{37}$	20,65	-7,40	17,45	-4,20	15,20	-1,95
36	O-*i*-Bu	20,70	-7,45	n.b.	n.b.	n.b.	n.b.
15	O-*n*-Bu	20,85	-7,60	17,60	-4,35	15,35	-2,10
16	O-*n*-Pr	20,90	-7,65	17,70	-4,45	15,45	-2,20
17	O-Et	21,60	-8,35	18,10	-4,85	15,60	-2,35
64	Et-CN	22,35	-9,10	n.b.	n.b.	n.b.	n.b.
33	S-*i*-Pr	22,45	-9,20	19,25	-6,00	16,55	-3,30
24	S-Et	23,10	-9,85	19,80	-6,55	16,70	-3,45
81	C(O)-*t*-Bu	23,10	-9,85	n.b.	n.b.	n.b.	n.b.
37	Se-Ph	23,20	-9,95	20,20	-6,95	16,85	-3,60
18	O-Me	24,10	-10,85	19,45	-6,20	16,30	-3,05
23	S-Me	24,30	-11,05	20,60	-7,35	17,10	-3,85
22	S-Ph	24,35	-11,10	21,25	-8,00	17,35	-4,10
65	C(O)-Ph	25,15	-11,90	n.b.	n.b.	n.b.	n.b.
66	O-*o*-Tolyl	29,05	-15,80	n.b.	n.b.	n.b.	n.b.
67	*o*-Ph-OPh	29,15	-15,90	n.b.	n.b.	n.b.	n.b.
19	O-Ph	30,20	-16,95	24,10	-10,85	18,95	-5,70
68	*o*-*t*-Bu-OPh	30,50	-17,25	n.b.	n.b.	n.b.	n.b.
69	p-Cl-OPh	33,10	-19,85	n.b.	n.b.	n.b.	n.b.
70	*p*-CN-OPh	37,85	-24,60	n.b.	n.b.	n.b.	n.b.
20	Cl	48,00	-34,75	36,40	-23,15	24,65	-11,40
71	F	54,70	-41,45	n.b.	n.b.	n.b.	n.b.

12 References

1. Haken H (1981) Erfolgsgeheimnisse der Natur. DVA, Stuttgart; (1984) The science of structure: Synergetics. Van Norstrand Reinhold, New York
2. Riedl R (1985) Die Strategie der Genesis, 4th edn. R. Piper, Munich
3. Weizsäcker CF von (1971) Die Einheit der Natur. Hauser, Munich
4. Woitschach M (1986) Gödel, Götzen und Computer. Horst Poller, Stuttgart
5. Jantsch E (1982) Die Selbstorganisation des Universums. dtv, Munich
6. Vester F (1980) Neuland des Denkens. DVA, Stuttgart
7. Hofstadter DR (1983) Gödel, Escher, Bach: An eternal golden braid, Penguin Books; (1985) Goedel, Escher, Bach: Ein Endloses Geflochtenes Band. Klett-Cotta, Stuttgart
8. Riedl R (1985) Die Spaltung des Weltbildes. Piper, Berlin
9. Bateson G (1983) Geist und Natur. Eine notwendige Einheit, 2nd edn. Suhrkamp, Munich; (1980) Mind and nature. A necessary unit, Bantam, Toronto
10. Breil H, Heimbach P, Kröner M, Müller H, Wilke G (1963) Makromolekulare Chem 69:18
11. Jolly PW, Wilke G (1974/75) The organic chemistry of nickel, vol 1 + 2. Academic Press, New York
12. Heimbach P (1973) Angew Chem 85: 1035; (1973) Angew Chem Int Ed Engl 12: 975; essential results of his own and results taken from several theses from coworkers of his team – Brenner W; Buchholz HA; Fleck W; Hey H-J; Meyer RV; Molin M; Ploner KJ; Scholz KH; Selbeck H; Thömel F; Wiese W – are described in [11]
13. Bartik T, Behler A, Heimbach P, Ndalut P, Sebastian J, Sturm H (1984) Z Naturforsch 39b: 1529
14. Kraushaar F (1986) Thesis. Universität-GH, Essen
15. Evans DA, Baillargean DJ, Nelson JV (1978) J Am Chem Soc 100:2242
16. Eigen M, Winkler R (1975) Das Spiel. R. Piper, Munich
17. Weyl H (1951) Symmetry. Princeton Press
18. Heimbach P, Kluth J, Schenkluhn H: KONTAKTE (Darmstadt) 1982 (2) 3
19. Heimbach P, Kluth J, Schenkluhn H: KONTAKTE (Darmstadt) 1982 (3) 33
20. Bartik T, Heimbach P, Schenkluhn H: KONTAKTE (Darmstadt) 1983 (1) 16
21. Bartik T, Heimbach P, Ndalut P, Preis H-G, Schenkluhn H, Sturm H: KONTAKTE (Darmstadt) 1983 (2) 14
22. Bartik T, Heimbach P, Schenkluhn H, Tani K: KONTAKTE (Darmstadt) 1984 (1) 44
23. Heimbach P, Bartik T, Szczendzina G, Zeppenfeld E: KONTAKTE (Darmstadt) 1988 (2) 45

24. Heimbach P, Bartik T, Drescher U, Gerdes I, Knott W, Rienäcker R, Schulte H-G, Tani K: KONTAKTE (Darmstadt) 1988 (3) 19
25. Traunmüller R, Polansky OE, Heimbach P, Wilke G (1969) Chem Phys Lett 3:300
26. Bartik T, Heimbach P: Proceedings 8th ICC Berlin 1984, V 573
27. Heimbach P, Bartik T, Boese R (1987) Phosphorus and Sulfur 30:155
28. Heimbach P, Bartik T, Boese R, Schenkluhn H, Szczendzina G, Zeppenfeld E (1988) Z Chem 28:121
29. Heimbach P, Bartik T, Preis H-G, Schenkluhn H, Szczendzina G, Zeppenfeld E (1988) Z Chem 28:237
30. Louven J-W (1984) Thesis. Universität-GH, Essen
31. Behler A (1985) Thesis. Universität-GH, Essen
32. Preis H-G (1983) Thesis. Universität-GH, Essen
33. Bestmann HJ, Zimmermann R (1982) Houben-Weyl, Methoden der Org Chem, vol E 1 p 715
34. Bestmann HJ (1980) Pure Appl Chem 52:771
35. Zeppenfeld E (1989) Thesis. Universität-GH, Essen
36. Enders D, Chemie Technik 1981: 504; Prof Dr W Winter, Grünenthal GmbH, gave us further information about the complexity of the problem:
    a) Fabro S, Smith RL, Williams RT (1967) Nature 215:296
    b) Scott WJ, Fradkin R, Wilson JG (1977) Teratology 16:333
37. Enders D, Hoffmann RW (1985) Chemie in unserer Zeit 19:177
38. Heimbach P (1974) Aspects Homogeneous Catal 2:81
39. Heimbach P, Schenkluhn H (1980) Top Curr Chem 92:45
40. Heimbach P, Schenkluhn H, Wisseroth K (1981) Pure Appl Chem 53:2419
41. Dewar MS, Dougherty RC (1975) The PMO-theory of organic chemistry. Plenum, New York
42. Heilbronner E, Bock H (1978) Das HMO-Modell und seine Anwendung, 2nd edn. Verlag Chemie, Weinheim
43. Fleming I (1976) Frontier orbitals and organic chemical reactions. John Wiley, London; (1976) Grenzorbitale und Reaktionen organischer Verbindungen. Verlag Chemie, Weinheim
44. Fukui K (1975) Theory of orientation and stereoselection. Springer Berlin Heidelberg New York
45. Epiotis ND et al. (1977) Structural theory of organic chemistry. Springer, Berlin Heidelberg New York
46. Seel F (1973) Grundlagen der analytischen Chemie, 5th edn. Verlag Chemie, Weinheim
47. Winkler-Oswatitsch R, Eigen M (1979) Angew Chem 91: 20; (1979) Angew Chem Int Ed Engl 18:20
48. Woodward RB, Hoffmann R (1972) Die Erhaltung der Orbitalsymmetrie. Verlag Chemie, Weinheim
49. a) Sustmann R, Trill A (1972) Angew Chem 84: 887; (1972) Angew Chem Int Ed Engl 11:838
    b) Sustmann R (1974) Pure Appl Chem 40:569
50. Heimbach P, Harsch G, Knott W, Tani K and Zeppenfeld E (to be published)
51. Hansch C, und Leo A (1979) Substituent constants for correlation analysis in chemistry and biology. John Wiley, New York
52. Seydel JK, Schaper KJ (1979) Chemische Struktur und biologische Aktivität von Wirkstoffen. Verlag Chemie, Weinheim

53. a) Lehmkuhl H, Keil T, Benn R, Rufinska A, Krüger C, Poplawska J, Bellenbaum M (1988) Chem Ber 121:1931
b) Keil T (1987) Thesis, Universität-GH, Essen contains a description of LFE-relations of the substituents X in the para-position of the phenyl group
54. a) Lehmkuhl H, Nehl H (1981) J Organometal Chem 221:131
b) Lehmkuhl H, Bergstein W, Henneberg D, Janssen E, Olbrysch O, Reinehr D, Schomburg G: Liebigs Ann Chem 1975:1176
55. Burger BJ, Santarsiero BD, Trimmer MS, Bercaw JE (1988) J Am Chem Soc 110:3134
56. Tolman CA (1977) Chem Rev 77:313
57. Levitt LS, Widing HF (1976) Progr Phys Org Chem 12:122
58. Litvinenkov LM, Popov AF, Gelbina ZP (1972) Dokl Chem 203:229
59. Bartik T, Himmler T, Schulte H-G, Seevogel K (1984) J of Organomet Chem 272:29
60. Manzer LE, Tolman CA (1975) J Am Chem Soc 97:1955
61. Schenkluhn H, Scheidt W, Weimann B, Zähres M (1979) Angew Chem 91:429' (1979) Angew Chem Int Ed Engl 18:401
62. a) Berger R, Schenkluhn H, Weimann B (1981) Transition Met Chem 6:272
b) Schenkluhn H, Berger R, Pittel B, Zähres M (1981) Transition Met Chem 6:277
c) Schenkluhn H, Bandmann H, Berger R, Gübinger E (1981) Transition Met Chem 6:287
d) Berger R (1980) Thesis. Universität-GH, Essen
63. a) Kluth J (1980) Thesis. Universität-GH, Essen
b) Brille F, Kluth J, Schenkluhn H (1979) J of Mol Cat 5:27
64. Schenkluhn H (1982) Habilitation. Universität-GH, Essen
65. a) Ferguson G, Roberts PJ, Aleja EG, Khan M (1978) Inorg Chem 17:2965
b) Aleya EC, Dias S, Ferguson G, Parvez M (1979) Inorg Chim Acta 37:45
66. a) Trogler WC, Marzilli LG (1974) J Am Chem Soc 96:7589
b) Trogler WC, Marzilli LG (1975) Inorg Chem 14:2942
c) Immirzi A, Musco A (1977) Inorg Chim Acta 25:L41
67. Bartik T, Himmler T (1985) J Organomet Chem 293:343
68. Gerdes H (1986) Thesis. Universität-GH, Essen
69. Glaubke V (1987) Thesis. Universität-GH, Essen
70. Schulte H-G (1983) Thesis. Universität-GH, Essen
71. Gerdes I (1986) Thesis. Universität-GH, Essen
72. Heimbach P, Kluth J, Schenkluhn H, Weimann B (1980) Angew Chem 92:569; (1980) Angew Chem Int Ed Engl 19:570
73. Rettig C (1987) Thesis. Universität-GH, Essen
74. Bock H (1977) Angew Chem 89:631; Angew (1977) Chem Int Ed Engl 16:613
75. Himmler T (1984) Thesis. Universität-GH, Essen
76. Bartik T, Heimbach P, Himmler T, Mynott R (1985) Angew Chem 97:345; (1985) Angew Chem Int Ed Engl 24:313
77. Tong et al. (1978) J Med Chem 21:732
78. Houk KN (1973) J Amer Chem Soc 95:4092
79. Houk KN (1975) Acc Chem Res 8:361
80. Fukuzumi S, Kochi JK (1981) J Am Chem Soc 103:7240
81. Woolley RG (1978) J Am Chem Soc 100:1073
82. West R (1987) Angew Chem 99:1241; (1987) Angew Chem Int Ed Engl 26:1231
83. Benson SW (1978) Angew Chem 90:868; (1978) Angew Chem Int Ed Engl 17:812
84. Burger K, Hein F (1979) Liebigs Ann Chem: 133

85. Hein F, Burger K, Firl J (1979) J Chem Soc Chem Comm: 792
86. Izumi Y (1983) Advan Catal 32:215
87. a) Reetz MT, Lecture, Universtität-GH Essen, Mai 1984
b) Steinbach R (1982) Thesis, University of Marburg,
c) Reetz M Th (1982) Top Curr Chem 106:1
88. Kaim W, Tesman H, Bock H (1980) Chem Ber 113:3221
89. Bock H, Kaim W, Rohwer HE (1978) Chem Ber 111:3573
90. Kaim W, Bock H (1978) Chem Ber 111:3843
91. Bock H, Wagner G, Kroner J (1972) Chem Ber 105:3850
92. Streets DG, Ceasar GP (1973) Mol Phys 26:1037
93. Schramm M, Thomas G, Towart R, Franckowiak G (1983) Nature 303:535
94. Heimbach P, Sunderbrink T (unpublished)
95. Heimbach P, Boese R, Glaubke V, Preis H-G (unpublished)
96. Vincent AT, Wheatley PJ: J Chem Soc Dalton Trans 1972:617
97. Reed FJ, Venanzi LM (1975) Helv Chim Acta 60:2804
98. a) Bachechi F, Zambonelli L, Venanzi LM (1975) Helv Chim Acta 60:2815
b) Bracher G, Grove DM, Venanzi LM, Bachechi F, Mura P, Zambonelli L (1980) Helv Chim Acta 63:2519
99. Sebald A, Wrackmeyer B, Theodoris CR, Jones W: J Chem Soc Dalton Trans 1984:747
100. Moore CE (1971) Atomic Energy Levels, NSRDS-NBS 35
101. Gilman H (1940) J Am Chem Soc 62:987
102. a) Metzner P: J Chem Soc Chem Comm 1982:335
b) Metzner P, Rakotonirina R (1983) Tetrahedron Lett. 24:4203
103. Normant JF, Alexakis A: Synthesis 1981:841
104. Krüger C (1980) Fresenius' Z Anal Chem 304:260
105. Goddard R (1980) Fresenius' Z Anal Chem 304:259
106. Theyssie P (1979) Fundam Res Homogenous Catal 3:107
107. a) Noe EA, Raban M (1975) J Am Chem Soc 97:5811 and literature cited therein
b) An excellent review of the literature was offered to us by Prof. E. Allenstein (†)
109. Becker G, Birkhahn M. Massa W, Uhl W (1980) Angew Chem 92:756; (1980) Angew Chem Int Ed Engl 19:741
110. Damewood Jr JR, Mislow K (1980) Monatshefte für Chemie 111:213 and literature cited therein
111. Linear and bent structures in "N"-systems seem to depend on alkyl/O-alkyl-substituents in 1,3-positions
a) Würthwein E-U (1984) J Org Chem 49:2971
b) Kupfer R, Würthwein E-U (1985) Tetrahedron Letters 26:3547
112. In the "P"-system the influences of substituents should be opposite.
113. a) Chetkina LA, Gol'der GA (1963) Kristallografya 8:194
b) Chetkina LA, Gol'der GA (1963) Kristallografya 8:582
114. Here we use the normal numbering of the main group elements: see Loening KL (1984) J Chem Ed 61:136
115. Spektrum der Wissenschaft, December 1985; Scientific American, December 1985
116. Kutzelnigg W (1984) Angew Chem 96:262; (1984) Angew Chem Int Ed Engl 23:272
117. Peoples PR, Grutzner JB (1980) J Amer Chem Soc 102: 4709
118. Lindsey AS, Jeskey H (1957) Chem Rev 57:583
119. Stork G, Boeckmann RK Jr (1973) J Amer Chem Soc 95:2014
120. Haenel MW, Lintner B, Benn R, Rufinska A, Schroth G (1986) Chem Ber 118:4922

121. Busse B, Weil KG (1981) Ber Bunsenges Phys Chem 85:309
122. Rees B, Heimbach P, Stede M (unpublished)
123. Schäfer HL, Gliemann G (1967) Einführung in die Ligandenfeldtheorie. Akademische Verlagsgesellschaft, Frankfurt am Main
124. Ohno K, Mitsuyasu T, Tsuji J: Tetrahedron Lett 1971:67
125. Molin M (1972) Thesis, Ruhr-Universität Bochum
126. a) Tsuji J (1979) Advan Organomet Chem 17:141
b) Tsuji J (1980) Organic synthesis with palladium compounds. Springer, Berlin Heidelberg New York
127. Arfsten N (1980) Thesis, Universität-GH, Essen
128. Zaar W (1974) Thesis, Universität-GH, Essen
129. Kelker M, Hatz R (1980) Handbook of liquid crystals, Verlag Chemie, Weinheim, p 43
130. Srinivasan R, Carlough KH (1967) J Amer Chem Soc 89:4932
131. Gleiter R, Sander W (1985) Angew Chem 97:575; (1985) Angew Chem Int Ed Engl 24:566
132. Malisch W, Blau H, Blank K, Krüger C, Liu LK (1985) J of Organom Chem 296:C32
133. a) Verhoeven JW (1980) Recueil Trav chim Pays-Bas 99:369
b) Verhoeven JW Pasman P, (1981) Tetrahedron (London) 37:943
134. Howard J, Waddington TC, Wright CJ, J Chem Soc Faraday Trans II 1976:513
135. About 90 percent of diethylether shows a zig-zag-arrangement: Jörgensen WL, Ibrahim H (1981) J Amer Chem Soc 103:3976
136. Brasanti P, Calo V, Lopez L, Marchese G, Naso F, Pesce G, J Chem Soc Chem Commun 1978:1085
137. Cahiez G, Normant JF, Bernard D (1975) J Organomet Chem 94:463
138. Tom Dieck H, Stamp L (1984) J Organomet Chem 277:297
139. Cardin DJ, Cetinkaya B, Doyle MJ, Lappert MF (1973) Chem Soc Rev 2:139
140. Cetinkaya B, King GH, Krishnamurthy SS, Lappert MI, Pedley JB: J Chem Soc Chem Commun 1971:1370
141. Schmidpeter A, Lochschmidt S, Willhalm A (1983) Angew Chem 95:561; (1983) Angew Chem Int Ed Engl 22:545
142. Schmidpeter A, Lochschmidt S, Sheldrick WS (1982) Angew Chem 94: 72; (1982) Angew Chem Int Ed Engl 21:63
143. Fukui K, Inagaki S (1975) J Am Chem Soc 97:4445
144. Brenner W, Heimbach P, Wilke G (1969) Liebigs Ann Chem 727:194
145. a) Thömel F (1970) Thesis, Ruhr-Universität Bochum
b) Heimbach P, Ploner K-J, Thömel F (1971) Angew Chem 83:285; (1971) Angew Chem Int Ed 10:276
146. Heimbach P (1973) J Synth Org Chem Jpn 31:299
147. Brenner W, Heimbach P: Liebigs Ann Chem 1975: 660
148. Heimbach P, Hugelin B, Peter H, Roloff A, Troxler E (1976) Angew Chem 88:29; (1976) Angew Chem Int Ed Engl 15:49
149. a) Roloff A (1976) Thesis, Universität-GH, Essen
b) Bandmann H, Heimbach P, Roloff A, J Chem Res (S) 1977: 261; J Chem Res Miniprint 1977:3056
150. Hey H-J (1969) Thesis, Ruhr-Universität Bochum
151. Wiese W (1972) Thesis, Ruhr-Universität Bochum
152. Noble WJ le (1974) Highlights of organic chemistry. Marcel Dekker, New York, p 402

153. Heimbach P, Tani K, Scheidt W, Twenty Fourth Symposium on Organomet Chem Japan 1976 (112–114), Preprints. The Chemical Society of Japan
154. Knott W (1988) Thesis, Universität-GH, Essen
155. Beez M, Bieri G, Bock H, Heilbronner E (1973) Helv Chim Acta 56:1028
156. Dauben WG, Kilbania Jr AJ (1971) J Amer Chem Soc 93:7345
157. Reißig HU (1984) Chemie in unserer Zeit 18:46, Fig. 2
158. Taft RW (1953) Steric effects in organic chemistry. John Wiley, New York
159. Meyer V (1894) Ber d chem Ges 27:510
160. Moore WJ, Hummel DO, Physikalische Chemie, 4th edn. Walter de Gruyter, Berlin, p 799
161. Krief A, Hevesi L (1984) Chimica Acta (Janssen) 2:3
162. Venet M (1985) Chimica Acta (Janssen) 3:18
163. Jonas K, Krüger C (1980) Angew Chem 92:513; (1980) Angew Chem Int Ed Engl 19:510
164. Helmchen G, Schmierer R (1981) Angew Chem 93:208; (1981) Angew Chem Int Ed Engl 20:205
165. Bähr W, Theobald H (1973) Organische Stereochemie, Heidelberger Taschenbücher, vol 131. Springer, Berlin Heidelberg New York, p 58
166. Stede M (1987) Thesis, Universität-GH, Essen
167. Ashby EC (1980) Pure Appl Chem 52:545
168. Budnik R, Heimbach P (unpublished)
169. Albery WJ, Knowles JR (1977) Angew Chem 89:295; (1977) Angew Chem Int Ed Engl 16:285
170. Masamune S, Choy W, Petersen JS, Sita LR (1985) Angew Chem. 97:1 (1985) Angew Chem Int Ed Engl 24:1
171. Nuhn P (1981) Chemie der Naturstoffe. VEB Akademie-Verlag, Berlin, p 445
172. Seebach D (1979) Angew Chem 91:259; (1979) Angew Chem Int Ed Engl 18:239 Despite the excellent examples of applying the concept of " Umpolung" there are a lot of problems still unsolved. So-called "tricks" in "alogical" connections are nothing more than normally applied processes in metal-catalyzed and – induced reactions. The question, whether the 1-, 3-, or 5- position or the 2-, 4- or 6-position in the alternating pattern of a π-system will react or if a (4q)- or a (4q + 2)- system will react preferentially, still remains unanswered
173. Bennett JF (1962) Angew Chem 74:731
174. Sicher J (1972) Angew Chem 84:177; (1972) Angew Chem Int Ed Engl 11:200
175. Saunders Jr WH, Fahrenholtz SR, Caress EA, Lowe JP, Schreiber M (1965) J Am Chem Soc 87:3401
176. Autorenkollektiv, (1977) Organikum. Deutscher Verlag der Wissenschaften, Berlin, p 282
177. Szczendzina G (1988) Thesis, Universität-GH, Essen
178. a) Brown HC, Moritani I (1956) J Am Chem Soc 78:2203
b) Schlosser M (1972) in: Methoden der organischen Chemie, Houben-Weyl, vol V/1b. Thieme, Stuttgart, p 134
179. Job P (1928) Ann Chim Paris 9:113
180. Wissing A, Heimbach P (unpublished)
181. Jones G (1967) Org React 15:204
182. Luft R, Basso J: Bull Soc Chim Fr 1967:986
183. Kryshtal' GV, Burshtein K Ya, Kul'ganek VV, Yanovskaya LA (1984) Izv Akad Nauk SSSR, Ser Khim 11:2541

184. Tani K, Noyori R, Otsuka S, et al. (1984) J Am Chem Soc 106:5108
185. Demuth M (1986) Angew Chem 98:1093; (1986) Angew Chem Int Ed Engl 25:994
186. Bestmann HJ (1986) Angew Chem 98:1007; (1986) Angew Chem Int Ed Engl 25:994
187. Berger R (1980) Thesis, Universität-GH, Essen
188. Bartik T, Gerdes I (1985) J Organomet Chem 291:253
189. Davies SG, Green MLH, Mingos DMP (1978) Tetrahedron [London] 84:3047
190. Davies SG (1982) Organotransition metal chemistry, Applications to Organic Synthesis. Pergamon, Oxford, Chap. 4 p 116
191. Bönnemann H (1985) Angew Chem 97:264; (1985) Angew Chem Int Ed Engl 24:248
Referring to Bönnemann the formation of "S/AS" products derived from the Co-catalyzed pyridine synthesis has to be seen related to ^{59}Co-shift data. By changing the sign of the ^{13}C-data belonging to the olefinic atoms of COD – i.e. exceeding the Cp• CoCOD as a "zero point" by change of symmetry in the structure – one gains a direct correlation between the product selectivity and the ^{13}C-shift data of the starting complex. Coherency (i.e. suppression of compensation phenomena) thereby guarantees the comparability of data.
192. Carbon dioxide as a source of carbon, Aresta M, Forti G (eds), (1988) Reidel, Dordrecht
193. The most important point is to avoid oxidation of the transitional metal complexes, e.g. by diprotonation etc.
194. Galli M, Trestianu S, Grob Jr K (1979) J High Resol Chrom 2:366
195. Piorr R (1982) Thesis, Universität-GH, Essen
196. Binger P, Büch M (1988) Top Curr Chem 135:77
197. Pettit R (private communication)
198. Boor Jr J (1963) J Polym Sci Part C 1:257
199. Su ACL, Colette JW (1975) J Organomet Chem 90:227
200. Brille F, Kluth J, Schenkluhn H (1981) J Mol Catal 5:27
201. Sisak A, Schenkluhn H, Heimbach P (1980) Acta Chim Acad Sci Hung 103:377
202. De Haan R, Dekker J (1976) J Catal 44:15
203. Bogdanovic B (1979) Adv Organomet Chem 17:105
204. a) Bogdanovic B, Henc B, Karmann HG, Nüssel HG, Walte D, Wilke G (1970) Ind Engl Chem 62:34
b) Wilke G, Verhandlungen der Gesellschaft Deutscher Naturforscher und Ärzte, 114. Versammlung, Munich 1986. Wissenschaftliche Verlagsgesellschaft, Stuttgart, p 275
205. Mende R (1983) Thesis, Universität-GH, Essen
206. Schlosser M (see references in [33])
207. Schmidbaur H, Stühler H, Buchner W (1973) Chem Ber 106:1238
208. Stock LM, Wright TL (1982) J Org Chem 47:597
209. Sturm H (1983) Thesis, Universität-GH, Essen
210. Padwa A (ed) (1984) 1,3-Dipolar cycloaddition chemistry, vols 1 and 2. John Wiley, New York
211. Wilke G (1988) Angew Chem 100: 189; Angew Chem Int Ed Engl 27:185
212. Deuchert K, Hünig S (1978) Angew Chem 90:927; (1978) Angew Chem Int Ed Engl 17:875
213. Bertalanffy L von (1949) Zu einer allgemeinen Systemlehre, Biol Gen 19:114
214. Huber RK (1980) Angew Systemanalyse 1:1
215. Halpern J (1983) Pure Appl Chem 55:99

216. Brown JM, Chaloner PA, J Chem Soc Chem Commun 1980:344
217. Brown JM, Murrer BA, J Chem Soc. Perkin Trans II 1982:489
218. Mittasch A, cited in Timm B, Proceedings of 8th International Congress on Catalysis, Berlin 1984, vol 1 p 18
219. Heimbach P, Kluth P, Schenkluhn H, Pullman B (eds) (1979) vol 12 p 227. Reidel, Dordrecht
220. Webster OW, Hertler WR, Sogah DY, Farnham WB, Rajan Babu TV (1983) J Am Chem Soc 105:5706
221. Jencks WP (1976) Accounts of Chemical Research 9:425
222. Fuson RC (1935) Chem Rev 16:1
223. Huisgen R, Dahmen A, Huber, H, Tetrahedron Lett 1969: 1461
224. Dahmen A, Huisgen R, Tetrahedron Lett 1959:1485
225. Heimbach P, Hey H (1970) Angew Chem 82:550; (1970) Angew Chem Int Ed Engl 9:528
226. Halevi EA (1976) Angew Chem 88:664; (1976) Angew Chem Int Ed Engl 15:593
227. Primas H, Müller-Herold U (1984) Elementare Quantenchemie. Teubner, Stuttgart
228. Blair J, Gasteiger J, Gillespie C, Gillespie PD, Ugi I (1974) Tetrahedron [London] 30:1845; Frequently the parity P = +1/-1 can be recognized with certainty, after faults of the *R/S*-nomenclature had been eliminated by introducing a factor N = +1/-1: Gasteiger J, GDCh-Fachgruppe Chemie-Information, Lecture, Würzburg, 17. March 1986
229. Hammond GS, de Boer CD (1964) J Am Chem Soc 86:899
230. Heimbach P (1974) Aspects of homogeneous catalysis, vol 2. Reidel, Dordrecht, p 81
231. a) Grob CA, Link H, Schiess PW (1963) Helv Chim Acta 46:483
b) Heimbach P (1964) Angew Chem 76:859; (1964) Angew Chem Int Ed Engl 3:702
232. Haken H (1978) Synergetics. Springer, Berlin Heidelberg New York
233. see [165] p 110
234. Reetz MT, Ostarek R, Pieko K-E, Arlt D, Bömer B (1986) Angew Chem 98:1116; (1986) Angew Chem Int Ed Engl 25:1108
235. Knabe J, Schamber L (1982) Archiv der Pharmazie 315:878
236. Bartik T, Heimbach P, Knott W (unpublished)
237. Kellogg RM (1984) Angew Chem 96:769; (1984) Angew Chem Int Ed Engl 23:782
238. Kessler H (1982) Angew 94:509; (1982) Angew Chem Int Ed Engl 21:512
239. Kessler H, Gehrke M, Haupt A, Klein M, Müller A (1986) Klin Wochenschr S7 64:74
240. Kessler H (1988) Angew Chem 100:507; (1988) Angew Chem Int Ed Engl 27:490
241. Frühbeis H, Klein R, Wallmeier H (1987) Angew Chem 100:413; (1987) Angew Chem Int Ed Engl 26:403
242. Heimbach AI, (unpublished)
243. Latest investigations show the high content of proteins in the bio-membranes
244. Bretscher MS, Spektrum der Wissenschaft, December 64, 1985; Scientific American, December 1985
245. Kaiser W, Ultrakurze Lichtimpulse und ihre Anwendung in Physik und Biologie, in Verhandlungen der Gesellschaft Deutscher Naturforscher und Ärzte, 114. Versammlung, München 1986, Wissenschaftliche Verlagsgesellschaft, Stuttgart, p 339
246. Michel H, Lecture, 7. Vortragstagung der Fachgruppe Biochemie der GDCh, 17. March 1988, Heidelberg

247. Hargittai I, Hargittai M (1986) Symmetry through the eyes of a chemist. Verlag Chemie, Weinheim
248. Mislow K, Siegel J (1984) J Am Chem Soc 106:3319
249. Cram DJ (1983) Science (Washington D.C.) 219:1177
250. Gutsche CD (1983) Acc Chem Res 16:161
251. Pöppel E (1985) Vom Vorteil des Vorurteils, in: Information und Kommunikation. Naturwissenschaftliche, Medizinische und Technische Aspekte. Wissenschaftliche Verlagsgesellschaft, Stuttgart, p 459
252. a) Corey EJ (1967) Pure Appl Chem 14:19
b) Corey EJ (1971) Quarterly Reviews, Chem Soc 25:455
253. Wisseroth K (1977) Monatsh Chem 108:141; (1978) Chem Ztg 102:45
254. Wisseroth K (1982) Chem Ztg 106:351
255. Primas H (1985) Chemie in unserer Zeit 19:109 and 160
256. Whitehead AN, Russell B (1910–1913) Principia Mathematica, 2nd edn. Cambridge University Press, Cambridge
257. Blumenfeld LA (1981) Biological physics. Springer, Berlin Heidelberg New York
258. Primas H (1983) Chemistry, quantum mechanics and reductionism, 2nd edn. Springer, Berlin Heidelberg New York

13 Epilogue
Nature, Life and Human Beings: Considerations of an Experimental Chemist

by PAUL HEIMBACH

Settlement of Chemistry

Due to the assignment of guilt by the media of a technical world arousing fears in man, chemistry is particularly burdened with the stain of alchemy, which in the public consciousness – overwhelmingly unfounded – in the mean occupation of quacks eager for profit and producing gold by deceit. Reference to this image leaves out totally the direct personal – even mental – relationship to matter and its transformation by the dedicated alchemist. Today the direct relationship between worker and matter is difficult to see. Certainly, this relationship is still present both in the work of a talented cabinet-maker and in the work of a creative artist both needing all their senses in order to make the correct choice of materials. We might assume this kind of devotion to matter as still being present with a modern biochemist, who strives for the infinite, sometimes artistic efforts in isolating enzymes as e.g. of photosynthesis [1] in a crystallized form and to opening up the way to numerous investigations. Therefore experimental chemistry is an applied art demanding unlimited devotion.

The transition from alchemy to chemistry over the centuries is a continuous one and perhaps it was introduced in its early beginning by Nikolaus Cancer de Coeße (Nikolaus von Kues or Nikolaus Cusanus: 1401–1464) postulating quantitative measurement [2] a long time before A.L. de Lavoisier (1743–1794). Chemistry is located between the pulls of physics and biosciences [3] and must stand this challenge. We are faced with statements of Bernard d'Espagnat [4] such as: "Evidentially physics is sufficiently developed now to claim the right to be accepted as the universal science of nature: 'nature' being obviously equivalent with reality in its right sense" or . . . "the basic principles of quantum mechanics do not rule the physics of the atom alone but the whole chemistry . . . and thereby any essential fact of the exact, empirical sciences." Being faced with such a claim to monopolize science put forward by a few physicists the statements of B. Russell [5] as philosopher offer great consolation to me as an experimentalist – and surely to many scientists representing other disciplines: "Physics is not mathematical because we have a lot of knowledge about the physical world but on the contrary because we know only a little bit we are able to recognize the mathematical properties of our world." "That is nice" physicist W. Heisenberg [6] would have said in addition.

Chemistry is in danger of losing young, inquisitive people who are attracted by the problems offered in biosciences. But chemistry might overcome its obvious, particularly selfmade crisis (see Chaps. 13.2 and 13.3) by being attractive to young people and being bound to and serving nature, life and mankind.

A wide bow stretches from mathematicized theoretical chemistry to experimental chemistry. I wish to account for the problem of an expanding collection of special disciplines (classical: inorganic, organic, physical and technical chemistry) in chemistry only briefly. On the other hand this diversification builds up a handicap for teaching and learning and therefore the call for a general chemistry grows stronger, giving the chance to offer a basic knowledge of related disciplines. The nearly complete banishment of experiment and intuition out of lectures will cause great damage to chemistry because the essential, simulated experience is left out. With reference to the problems of combining aspects of forming and formed information from alternative principles in chemistry should be pointed out, in order to elucidate the extreme importance of complementary thinking. Theoreticians like H. Primas [8] would surely take to mathematically secured ways to realize this aim. He referred to the motto of the physicist C.A. Truesdell [9] cited in the summary: "It is a pure illusion to think that learning all about tiny things is the path to knowledge about big things."

Being an experimentalist I have to try another way (small is beautiful). Nevertheless H. Primas [10] says: "Phenomenological models are of great importance to both engineer and experimental scientist and they should never be undervalued. In most cases they combine empirical laws and deeper theoretical insights in an artistic manner. Each model makes use of non-formalized and unreflected theoretical understanding and thereby it should be used only in combination with a lot of 'common sense' ". If the experimentalist is determined not to lose his 'common sense', he has to approve of every epistemological aid being offered. Possible ways leading out of a crisis by means of philosophical questions, too, will be discussed by a few examples.

The tasks of an experimental chemist are widespread from the preparation of highly purified silicon used in the production e.g. of integrated circuits up to the evolution of catalytical processes for the preparation of highly complex molecular and polymer substances or – as mentioned above – up to the isolation of very sensitive enzymes in pure form. Looking at chemical catalysis we have still to accept the confession of the catalysis scientist A. Mittasch [11]: "The multicomponent catalyst is the winning catalyst!" Or to say it in other words as I did: "A lot of help helps a lot! The problem remains: What kind of help? How much?" The experimentalist engaged in developing catalytical processes is faced with a paradox: by adding more and more substances to another substance the resulting mixture becomes increasingly inhomogeneous. In contrast a chemical process – as an event of new qualities of substances to come into existence – might become more and more selective by adding more substances – in form of effectors and catalysts – having the right properties and an adequate amount. The "obtaining" of new, unitary qualities represents the fascination of chemistry and can result only from the harmonious cooperation of many small effects (small is beautiful).

A Crisis in Chemistry and a Possible Way to Overcome It

Despite euphoric successes of the chemical industry in world trade – or perhaps provoked thereby – chemistry as a science is undergoing a crisis [13]. The active and investigating human being shifts from the center of interest to the border and science becomes sterile [14]. The time is ripe for releasing movement and development from the current situation. The common uneasiness should only raise new questions not mislead to laziness or resignation. The rejection of the guardians who uphold the ruling paradigms [15] will surely be necessary to shorten the long way between intuitive experiences and the clear cut, unequivocal rules (rules/inverse rules) by perseverance and make the formulation of new epistemological truth possible. A correct analysis of the new insight and an integrating abstraction are only achieved by overcoming the exaggerations of intuition primarily to be seen in a lack of clearness and precision in detail and a broadening of new terms. As a teaching experimental scientist one should submit oneself to the task e.g. of adding the corresponding inverse rules to the newly formulated rules whenever possible, in order to define the borderline to the still "underscribable" much easier. The different terms, rules and – to many – exceptions frequently do not allow any insight into limitations and lead to one-sided considerations not being necessary. By use and abuse of one-sided or latent postulates and rules the foundation of a pseudoscience has to be seen as a danger. As mentioned above and accepted as a necessity Th. S. Kuhn [15] states "the trials to publish new insights being ignored or being judged as meaningless". This fact should not be discouraging because R. Spaemann [16] connotes that the Hiob-messenger of today is not killed but neglected totally and his message is defined as being irrelevant. On the other hand the scientist, who reveals a paradigm as a partial truth, has to give an acceptable form to his insight, i.e. creating hope and relief for tomorrow. As human beings we need hope for tomorrow and not a condemnation of past and present.

Overcoming the crisis is to be strived for by simplification. "But nevertheless, accepting simplification, the essential point must be repeated correctly" [10]. This task is hardly to fulfill as demonstrated by the remark made by the aphorist G. Elgozy [17]: "Complicated: Nothing is more difficult than simplification. Nothing is simpler than complication". From an inspiration of my wife I wish to cite two aphorisms of St. J. Lec [19] to illustrate that the "right way" itself might lead in "different" directions. Lec enumerates the possibilities offered by a river as a dynamic mean of transport:

(1) "To reach the source one has to swim against the stream."
(2) "The clever guy swam following the stream – bravely – and was drowned. The even more clever guys reached their aim – by following the banks:"

Perhaps it would be better to describe the latter case in other terms:

The knowing guy swam following the stream – bravely – and was drowned.
The wise men reached their aim – following the banks.

This demonstrates the fact that theoreticians and experimentalists obviously choose divergent ways. The physicist M. Planck described the efforts of theoreticians to reach the sources of knowledge in his lecture "Vom Relativen zum Absoluten" in an excellent way [20]. Practitioners, goal orientated, have to go the way "Vom Absoluten zum Relativen" (see Chap. 3) to get rules for the course of action. Every effort following both directions is worth while. Fascinating aspects in systems of increasing complexity can be solved [21]; e.g.:

(1) Finding of new, latent qualities and causalities by means of pattern recognition.
(2) The discovery of new, unusual properties of materials (advanced materials) by the cooperation of smallest effects (in properties and amounts).
(3) Recognition of multicausal networks with causal cross connections.
(4) Cooperative effects of smallest variations leading to unity.
(5) Searching for information destroying compensations.
(6) The gain in information by using possible compensations.
(7) Overcoming of obviously experienced, practiced programs by multifold methods (adaption to the complexity of systems).
(8) Learning the art of asking 'correctly' in complex systems [22].

Problems Centered on Quantity

An undoubted factor influencing the acceptance of a publication in a journal of chemistry is still to be seen in the fact that one or more new compounds are described. However, this takes place in spite of there being about 7 million known chemical compounds and their number increases practically in an exponential way. By restricting oneself to the extent of the contents of Chemical Abstracts for 5 years only, one might anticipate it's "ironical value – but ironical in its very strict sense" [23]. The broadened, formalized quantity of fixed, separated facts determines the development of young chemists instead of an integrating power based upon the insight in ordering and ruling, cooperating factors and the fascination arising from unanswered questions. A student wishing to complete his studies within a reasonable time has to stay in the laboratory during the day or has to submit himself to lectures and seminars giving facts and has to write his records at night. It is a rather misanthropic study. Creativity and delight springing from alternative occupations are banished to a large extent. My efforts are directed at a change in the system.

Another problem of common epistemological interest will be discussed here in more detail. The concept of molecular architecture is an procedure of highly ranked, heuristical value to experimentalists i.e. the separation of a molecule (as an unit) into subsystems (parts) considered to be of importance. Functional groups (e.g. acid groups) and parent systems (e.g. benzene) and different substituents (influencing e.g. the behavior of the functional group) are of great importance in experimental chemistry (see also the general formulation in Fig. 13.1).

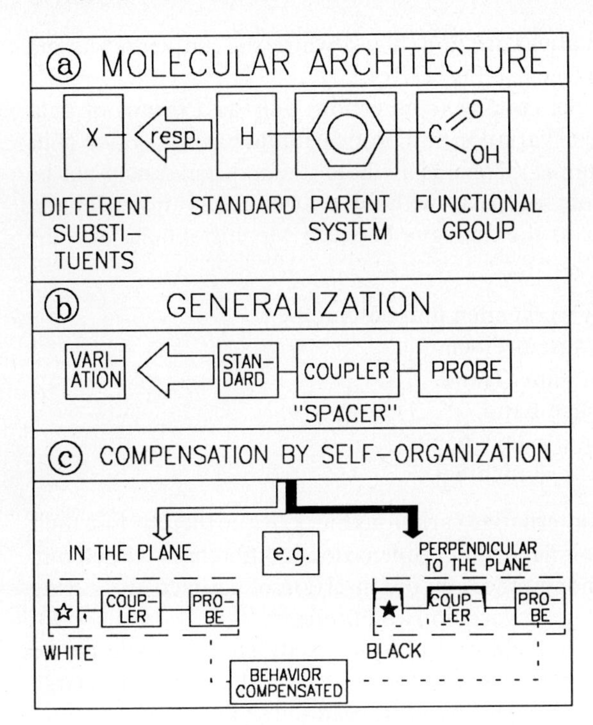

Fig. 13.1a-c. The problems of separation into parts in chemical **(a)** and in general systems **(b)** are depicted; in **(c)** compensations by self-organizations of the whole system are demonstrated by example

Quantitative structure-reactivity-relationships are highly estimated in chemistry [24,25]; but in some ways they lead us up a blind alley. To shift from assertion to example we choose the simple system of Fig. 13.1 in a general form.

The "one or the other" behavior of the probe (chemically: thermodynamic and kinetic behavior of the functional group) is investigated by experiments and the variations (of substituents) referring to a standard (in most cases hydrogen as substituent) can be described mathematically in form of so-called Linear-Free-Energy-Relationships (LFE) following L.P. Hammett's proposal. Experience gained by experiments is substituted by a constant algorithm and variations are obviously taken into consideration by parameters (quantities) [26]. Indeed, if one introduces, with respect to new probes (other functional groups), new parameters (for the type of reaction), one might succeed in the prediction of the behavior of new probes supported by the well-known parameters of variations (substituents) and a slightly modified algorithm (mathematical rule).

Nevertheless mathematical formulations represent an abstraction of reality and therefore the same quantity can describe different structures in principle (same parameter of different variations e.g. alternative black/white, i.e. chemically different substituents like $-OCH_3$/$-SCH_3$ bearing oxygen or sulphur as the central atom). This can easily be seen in the presence of information destroying compensation phenomena (if they are allowed to arrange themselves freely). Released by an alternative variation (black/white) the coupler reacts and takes its alternative position (see e.g. Fig. 13.1c).

The idea of looking at the coupler (connection) only as a means of guaranteeing the distance between variation and probe without having a value of its own is misleading. The dynamic system as a whole (the molecule) may not be described as an arrangement of parts without critical reservation. Johann Wolfgang von Goethe gives us the dialogue between Mephistopheles and the student in Faust:

> "Wer will was Lebendig's erkennen und beschreiben,
> sucht erst den Geist herauszutreiben.
> Dann hat er die Teile in seiner Hand,
> fehlt leider nur das geistige Band.
> Encheiresin naturae nennt's die Chemie.
> Spottet ihrer selbst und weiß nicht wie".

The same parameters of alternative variations black/white thereby lose their meaning completely, if the coupler – the compensator of differences – is left out. We could prove this fact among others by the insertion of couplers displaying alternative arrangements between variation and probe.

The dramatically different behavior of systems by choosing alternative variations black/white was synchronized by compensation phenomena to a large extent. Even LFE-relationships might therefore lead only to parts of the truth. They do not meet the demands of describing the Moiré phenomena to be expected in chemical systems.

We can account for a possible dilemma of quantities: they do not allow us to draw inference from changing qualities often compensated by each other. T. Bartik, my former coworker, pointed a way out by relating quantities under coherent conditions by the choice of a degenerated standard and by arranging the quantities in patterns. Idealized one can expect five pairs of inverse patterns principally by exchange of three identical variations by three others at a defined center with at least one (three, five, . . .) quality(ies) to be changed in order to get an inverse pattern (see Fig. 13.2). On the other hand we succeeded in casting light on some latent qualities (see Chap. 3) by a convenient choice of standard. Classical kinetics as a very successful method for comprehending relationships quantitatively – gives us another example for the failure of an approved reductionistic method in complex chemical systems (see also the analysis of efficiency in enzymes [28] and especially the interesting considerations of L.A. Blumenfeld concerning the internal pool of heat in a given system [29]).

We have accepted quickly formalized structures in thinking as depicted in Fig. 13.3 according to G. Bateson [18] in their relevance to solving open problems. We collected the essential terms on the separated levels of increasing complexity i.e. for systems (individuum/class/meta-class), for models (digital/analogous/digital and analogous) and for methods (counting/measuring/pattern comparison).

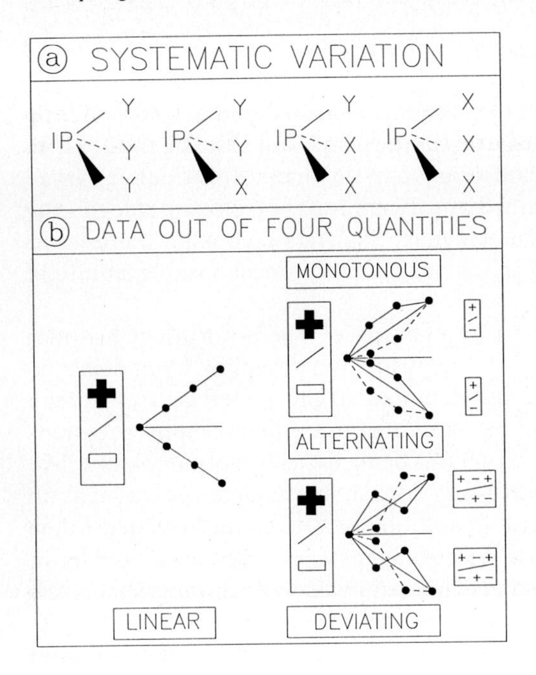

Fig. 13.2a,b. Five pairs of alternative patterns consisting of three data and a standard [27]

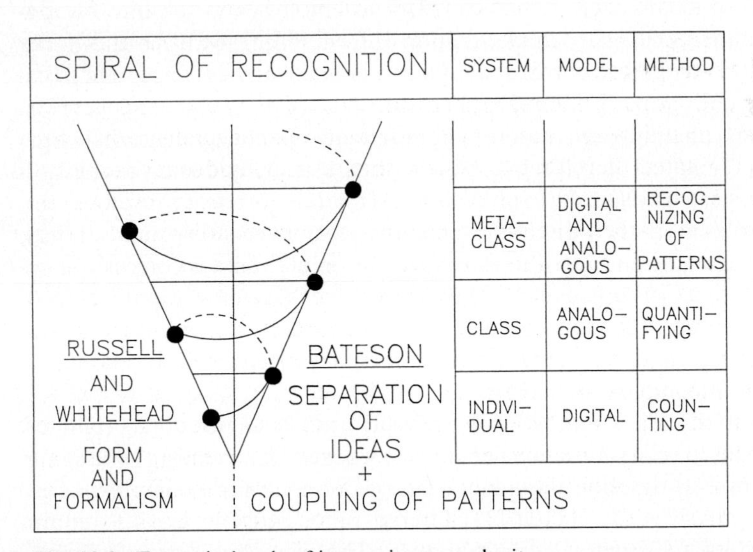

Fig. 13.3. Terms in levels of increasing complexity

Problems Centered on the Language of Chemists

"The history of science is a history of revealing one-sided partial truths offering a refuge because they have been regarded as the whole and absolute truth" (J.W. Hayward [30]). This emphasis laid upon one-sidedness holds out far more obvious security than that to be gained by a dynamically developing recognition. W. Porzig stated additionally: "The language transfers each kind of non-illustrative relationships into terms of space. . . . This is an unchangeable feature of human language.

Relations in time are expressed in terms of space: – before or after Christmas –, – within a period of two years –". Chemistry as a discipline is not detached from other sciences and the chemist is bound to the preferences of human language (see [31]). We should not be astonished to see that complex situations might give rise to the unreflected use of terms being basically not comparable like "steric" (derived from the alternatives key/lock in mechanics), once used as an obviously alternative and otherwise in addition to "electronic" (related to the alternatives repulsion/attraction of electromagnetic interactions). "Errors in logical typing" can arise by connection of non-equivalent terms (see [18]). For the necessary, detailed discussion see Chap. 3.8, Fig. 3.23 and context.

I wish to point out an aspect of major importance which should be kept in mind permanently. To draw a veil over the little difference or to select the incorrect terms and alternatives in a given situation can be equivalent to a falsification of the truth or with a transmission of only the partial truth.

Complexity in chemistry is thereby accompanied with non-illustrative facts accounting for the one-sided use or the misleading relation of general terms. How to overcome these terms or how to detect their narrow range of validity can be realized easier by means of epistemology [18]. Epistemology and "philosophy can perhaps be defined as the perpetual intention of formulating the basic questions. Modern science recognized correctly that it can not take any advancement from philosophy in daily work but it might get disturbed". (C.F. von Weizsäcker [7]). Periods of crisis (including the change of paradigmata) do not belong to daily life. Th.S. Kuhn [15] stated therefore: "I believe that, especially during periods of established crisis, scientists take to philosophical analysis as being convenient for solving the problems of their discipline. Scientists do not need to be philosophers normally and are not committed to changing this habit". That is correct!

"The Ordering Concept of the Basis of Alternative Principles" was developed by me together with young, enthusiastic coworkers and friends in order to gain practical, alternative rules leading the experimental procedure in the evolution of chemical multicomponent systems.

Fundamentally we intend to comprehend and to control the origin of alternative order in relatively simple chemical systems. The chemist I. Prigogine [32] was the first to describe the generation of order in more complex, open systems. The physicist H. Haken [33] offered other possible ways to attain understanding of self-organization in matter. The way chosen by an experimentalist is much more naive.

From Absolute to Relative

First of all I wish to illustrate the roots of the Concept Of Alternative Principles localized in the science of chemistry by means of simple examples. Chemists are basically familiar with alternative order e.g. in space (see Fig. 13.4a):

(1) OBJECT/MIRROR IMAGE are equivalent in energy (degenerated and thereby paritetic) and
(2) e.g. *cis-/trans*-arrangements are different in energy (and thereby complementary).

This order can remain or it can be inverted during a chemical process as to be seen in Fig. 13.4b in the case of a spatial uniform molecule. Referring to the model

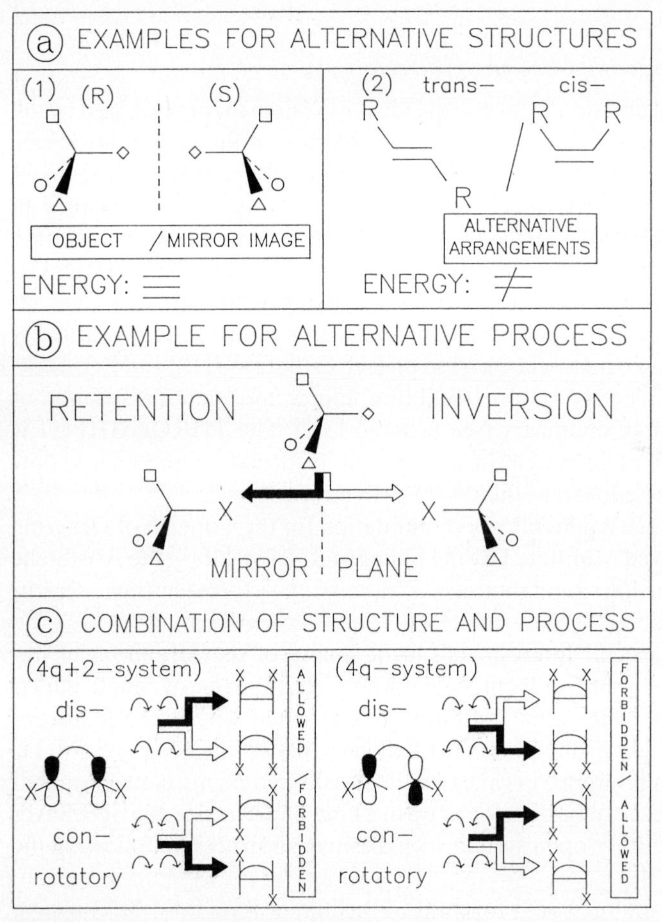

Fig. 13.4a. One example for each paritetic and complementary alternatives. **(b)** Retention and inversion of a paritetic order

of electromagnetic interactions the process might finish up in a simple case (with retention of phase relations during process: conservation of orbital symmetry [34]) in a clockwise of anti-clockwise sense parallel/antiparallel (con-/disrotatory) directed by the phase conditions of the system in α,ω-position (Fig. 13.4c). Only the known position of two spatially defined subsystems (*cis-/trans-*) and the known spatial arrangement (*cis-/trans-*) in the product allow in combination to decide experimentally what kind of process has taken place. We are not enabled to judge a clockwise or an anti-clockwise movement here. Beside the absolute case (100% of one or the other arrangement: yes/no decision) reality demonstrates relative decisions in a chemical experiment more frequently (activated/inhibited) and symmetry inversions e.g. from cis- to trans-arrangements. A decision whether little alternative effects will cooperate in one or the other direction (differentiation) or, whether they will show compensation phenomena, is only realized in the case of at least 2×2 variations.

Foundation of the Ordering Concept of Alternative Principles

It was a fundamental decision to take to the ideas of system theory initiated by the biologist L.V. Bertalanffy [35], i.e. to accept a division of the system into three parts IN-/THROUGH-/OUTPUT and to leave the usual questions of chemistry behind, concerning reaction mechanisms (the questions of the structure in internal dynamics: THROUGHPUT) and to shift to the transfer of information between IN- and OUTPUT (external dynamic). Stimulated by ideas of H. Haken [33] concerning self-organization in matter we searched for the important alternatives in the INPUT, looking at chemical systems of increasing complexity from individuum/class/meta-class and so on (see G. Bateson [18] and Fig. 13.3), i.e. in the valence electrons/atoms/building stones/molecules/associates of molecules and so on, of chemistry (see Scheme 13.1). The THROUGHPUT is regarded as being an open system in which mass, information and energy flow from the INPUT (the educts) to the OUTPUT (the products).

In order to arrive at a general, good foundation for the Concept of Ordering in chemistry, I followed – initiated by the biologist R. Riedl [36] – the Aristotelic idea and arranged the four good roots in a complementary form (causa materialis and efficiens upward and causa formalis and finalis downward).

It might be of special interest here to be aware of the formation of unequivocal units (holographs) from cooperating alternatives of small effects following causa efficiens along the multidual decision trees. The typical feature of a holograph to spotlight, is the presence of all information in each part of it, i.e. a network unit. With reference to causa finalis I wish to stress my conviction that every chemical process can be developed up to a kind of perfection that it behaves like a "Mode" (an unitary open system with dissipating structure) following the idea of H. Haken [33].

One can infer from the causa formalis of the higher level how the choice of building stones with increasing complexity released the effect wanted.

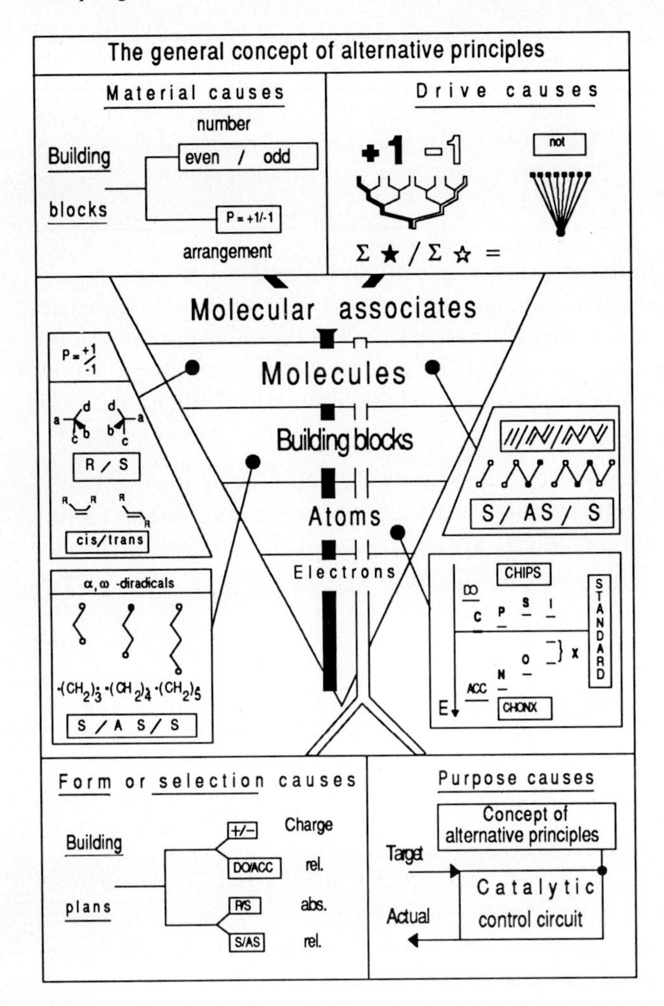

Scheme 13.1

From Alternatives to Complementarities

A very important decision for life is already done on a molecular level. The construction of biological systems is only based upon building stones of one parity (object or mirror image), e.g. 'D'-sugars and 'L'-aminoacids. A new aspect of so-called secondary qualities [23] of chemical substance must be taken into consideration. OBJECT/MIRROR IMAGE (the enantiomers) of a collection of identical molecules (substances of one parity) display different effects in biosystems of only one parity, e.g. [37]:

Effect	object	/ mirror image
taste	bitter	/ sweet
odour	lemon-like	/ orange-like
pharmaceutical	extreme teratogenic (thalidomide)	/ sopophoric

These effects depend on the so-called diastereomeric interactions.

My eldest daughter Almut pointed out some very interesting consequences arising from this unequivocal parity realized in the molecular building stones from her point of view as a physician (1987) in a yet unpublished investigation: "The relevance of the ordering Concept of Alternative Principles for a few biomolecules".

(1) Double layers established by biomolecules of one parity in membranes consist of two layers, which do not behave as object and mirror image (see Fig. 13.5.a). Not the building stones are different but the kind of interaction inside/outside via interactions with an α-helix of proteins.
(2) The right and the left hand are – at least on a molecular level – not object and mirror image (see Fig. 13.5.b), therefore differentiating reactions by the diastereomeric interactions with spatially defined molecules in the right and the left part of body can be avoided. That is the reason for abolishing the misleading term chirality and adopting the use of the term parity in chemistry.
(3) Just to see so the two parts of the brain cannot be paritetic in principle but complementary.

G. Bateson [18] has discussed the special complexity of the pair right/left in detail (people say: right is there, where the thumb is left).

Perspective

One idea is surely fascinating and can be seen directly: based upon the early decision to restrict to only one parity on the molecular level for life, e.g. at the frontiers of decision (membranes), one might infer that this decision is maintained through all levels of increasing complexity and that our thinking, too, is ruled overwhelmingly by complementarities and only to a small extent by really paritetic alternatives. Does this idea perhaps reflect one of the deeper meanings of the intuitively chosen, old far-eastern symbol called T'ai-chi T'-u (Fig. 13.6), perhaps warning to overestimate the direct decisions of yes/no (either/or) and to accept the unit of (as well/as)-decisions in their deservedly, overwhelming part? This should perhaps be valid for a lot of decisions, erroneously taken to be of a contrary type like:

inside	/	outside
woman	/	man

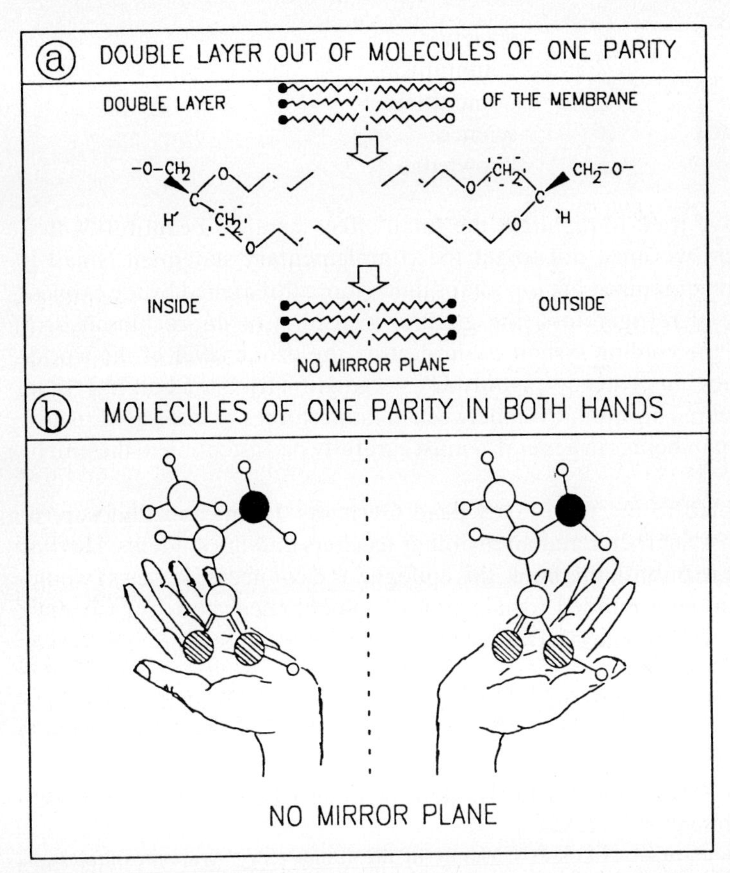

Fig. 13.5a. Membranes as two dimensional liquid crystals consisting of building stones of one parity have no mirror plane between the layers. **b** Hands are formed by building stones of one parity

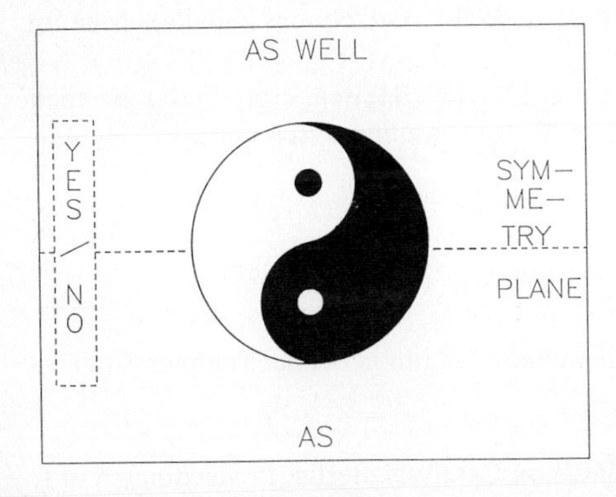

Fig. 13.6. One possible interpretation of a far-eastern symbol

individual	/	society
capitalism	/	communism
euphoria	/	admonition
social science	/	science
belief	/	knowledge

Whenever I have tried to highlight the small effect (small is beautiful) with a special intention we must not forget the complementary statement (small is horrible). Highly organized life is perhaps much more threatened by the careless throwing away of refrigerators (the gradual vanishing of fluorochlorohydrocarbons out of the cooling system could destroy the ozone layer of the world, which guarantees our protection against UV – radiation from the universe) than by the usually feared big atomic war between the superforces. Social scientists and scientists – human beings in general – must carefully pay attention to the unit in nature.

I am indebted to my family, my dead German-, Latin- and Philosophy-teacher Dr. Ernst Schroeter and all my other teachers and my students. Having the opportunity to publish this book, this epilogue and connected reviews I would like to thank an open-minded female editor S. Stiehl (Zeitschrift für Chemie, GDR), Dr. R. Klink (Kontakte, Darmstadt) and Dr. R. Stumpe (Springer-Verlag Heidelberg).

References and Notes

1. Deisenhofer J, Huber R, Michel H: Nobelpreisträger für Chemie 1988; (1988) Chemie in unserer Zeit 22:A53
2. In the foundation of Nikolaus Cusanus in the monastery "Armenhospital zum Heiligen Nikolaus von Kues" in Kues on the banks of the Moselle is his large private library. There, 340 handwritten manuscripts from the 9th to the 15th century can still be seen.
3. Heimbach P, Kluth J, Schenkluhn H: KONTAKTE (Darmstadt) 1983 (1) 16
4. D'Espagnat B (1983) Auf der Suche nach dem Wirklichen, Aus der Sicht eines Physikers, Übersetzung A. Ehlers. Springer, Berlin Heidelberg New York
5. B. Russell's statement is cited in: Köstler A (1980) Wurzeln des Zufalls suhrkamp taschenbuch 181, 3rd edn, p 101
6. The following situation is shown in [7] p 126, 127: Heisenberg: "Natur ist eben mathematisch einfach." Frage: "Was heißt denn mathematisch einfach?" Heisenberg: "Das ist eben schön".
7. Weizsäcker CF von (1974) Die Einheit der Natur. Deutscher Taschenbuch Verlag, Wissenschaftliche Reihe, München
8. Primas H (1982) Chemistry and Complementarity, Chimia 36:239
9. Truesdell CA, see [1] in [8]
10. Primas H, Müller-Herold U (1984) Elementare Quantenchemie. Teubner Studienbücher Chemie, Stuttgart
11. Mittasch A cited in [12]
12. Timm B (1984) 8th International Congress on Catalysis, Berlin, Proceedings, Vol 1, p 18

13. Primas H always points out that also in theoretical chemistry exclusive thinking about molecules must be overcome.
14. Hoffmann R (1988) "Die chemische Veröffentlichung – Entwicklung oder Erstarrung im Rituellen?", Angew Chem 100:1653
15. Kuhn TS (1976) Die Struktur wissenschaftlicher Revolutionen, suhrkamp taschenbuch wissenschaft 25, 2nd edn. Suhrkamp, Frankfurt
16. Spaemann R (1987) Das Natürliche und das Vernünftige, Aufsätze zur Anthropologie. Serie Piper, München-Zürich
17. Elgozy G, cited according to [18]
18. Bateson G (1983) Geist und Natur. Eine notwendige Einheit, Suhrkamp, München, 2nd edn; (1980) Mind and Nature. A Necessary Unit. Bantam Books, Toronto New York London Sydney
19. Lec St J (1) (1981) Neue unfrisierte Gedanken, Carl Hauser, München, 10th edn, p 29; (2) (1986) Steckbriefe, Carl Hauser, p 11
20. Planck M, Vom Relativen zum Absoluten, Gesammelte Abhandlungen und Vorträge, Vol 3, Lecture 1. Dezember 1924
21. Koch MG, Spektrum Der Wissenschaft 1986: (10), 12
22. Crick FHC, "Gedanken über das Gehirn", Spektrum der Wissenschaft 1979:147; Scientific American 241:181 (1979)
23. Zollinger H (1988) Chemie – Teil eines Ganzen, SWISS BIOTECH 2:13
24. Hansch C, Leo A (1979) Substituent constants for correlation analysis in chemistry and biology. John Wiley, New York
25. Seydel JK, Schaper KJ (1979) Chemische Struktur und biologische Aktivität von Wirkstoffen. Verlag Chemie, Weinheim
26. With a fixed algorism (calculation procedure) characteristic values are formed dependent on the properties; unfortunately these quantities are often misleadingly called constants.
27. As the standard, triphenylphosphane with 3 phenyl groups is used which adopts 2 energy degenerate forms.
28. Albery WJ, Knowles JR (1977) Angew Chem 89:295; (1977) Angew Chem Int Ed Engl 16:285
29. Blumenfeld LA (1981) Biological Physics. Springer, Berlin Heidelberg New York
30. Hayward JW (1986) Der Zauber der Alltagswelt, Ein tieferes Verständnis der Wirklichkeit durch Wissenschaft und Weisheit. Droemersche Verlagsanstalt Th. Knaur Nachf, München
31. Porzig W (1971) Das Wunder der Sprache. Francke, München Bern
32. Prigogine I, Stengers I (1981) Dialog mit der Natur, Neue Wege naturwissenschaftlichen Denkens. R Piper, München-Zürich
33. Haken H (1981) Erfolgsgeheimnisse der Natur, DVA, Stuttgart; (1984) The science of structure: Synergetics. Van Norstrand Reinhold, New York
34. Woodward RB, Hoffmann R (1972) Die Erhaltung der Orbitalsymmetrie. Verlag Chemie, Weinheim
35. Bertalanffy L v (1949) Zu einer allgemeinen Systemlehre, Biol Gen 19:114
36. Riedl R (1985) Die Spaltung des Weltbildes. Piper, Berlin
37. Enders D, Chem Tech 1981:504; Prof Dr W Winter, Grünenthal GmbH, gave us further information concerning the complexity of the Contergan (Thalidomide) problem:
   a) Fabro S, Smith RL, Williams RT (1967) Nature 215:296
   b) Scott WJ, Fradkin R, Wilson JG (1977) Teratology 16:333

14 Subject Index

ABSOLUTE/RELATIVE 68, 69
activated (see glossary)
activation/inhibition 8, 27, 115
additivity rule 10
agonist/antagonist 35, 87, 164, 170, 171
alkali metals compared 43, 45, 86, 88
alkyl series
– Me/Et/i-Pr/t-Bu, branched 15, 17, 29, 31–33, 76–78
– Me/Et/n-Pr/n-Bu, homologuous 27, 77
– H/Me/i-Bu/i-Pr, biomimetic 45, 77, 78, 79
alkyl-/alkoxi compared 18, 20, 99, 100
allene isomerism 51
allophasic (see glossary)
allosteric (see glossary)
alternating phenomena (see glossary) 17, 27, 31, 33, 50–52, 77, 78, 86, 101
alternative positions 77
– α-/β- 29, 61, 102
– aldehyde/ylide in a Wittigsystem 75
– in/at the system 57, 162
– 1-/2- or 1, 4-/2, 3- in dienes 61, 62, 77, 90
– vinylic/allylic 61
alternative principles (see glossary)
– definition 150, 151

BERNOULLI apparatus 71
biomembranes 43, 170, 171
biomimetics 6, 29, 77, 141, 163
BIRCH reduction 22, 23

catalysis model 177
^{13}C-NMR-data 21, 34, 98
CHIPS-chemistry 38, 66
chirality (see glossary)
CHOCK-chemistry 41
CHOICE-chemistry 41
CHONX-chemistry 38, 66
cis-/trans-decisions 27, 40, 43, 51, 63, 94
Cl, Br/I compared 40, 79, 80, 86
compensation phenomena (see glossary) 27, 36, 72, 86, 158, 163
complementary alternatives (see glossary) 158–160, 167, 170–173, 206
complex chemical systems 1, 116, 137, 179
concentration control maps in 113, 116
– asymmetric syntheses 96, 99, 101, 102
– metal catalysis 118, 120, 121, 127, 129
– nitration by Hg^{++} 127
– organic reactions 123, 126
con-/dis-rotation 6, 36, 147, 203
conformational changes 26, 36, 40, 178
constants (see glossary)
co-oligomerization of 139
– butadiene and alkynes 48, 58, 59
– butadiene and azines 59

– butadiene and olefine 12, 59, 120
– butadiene and oxo-compounds 46, 59, 92
– butadiene and Schiff'bases 46, 139
COPE rearrangement 2, 155, 158
COVALENT/IONIC 63, 65
C/Si compared 17, 18, 20–22, 27
Cu-reagents 39, 56
CYCLIC/OPEN structure 2, 31, 46, 52, 54, 56, 75, 84, 158, 163

Δ1-/Δ2-olefines 6, 84–87, 126
diastereomeric interactions (see glossary)
DIECKMANN cyclization 45, 74, 84
DIELS-ALDER reaction 6, 22, 59, 139, 141
differentiation (see glossary)
digital decision trees 6, 15, 63, 81, 84, 102
dimerization (see oligomerization)
direct splitting 60, 161–163
direct unifying 54, 55, 60, 102, 142, 145, 151, 161–163
DNS 167
DO/ACC heteroatoms 2, 18, 37–40, 49, 129, 131
"DO/ACC" substituents (see glossary) 2, 29, 32–34, 49
double dual decisions 65, 72, 75, 80, 154, 155, 159, 160, 162, 174, 178
dynamic standard 69

eight-/six-membered rings 5, 20, 50
electronic (see glossary)
enamine synthesis 94
entropy (see glossary)
END ON/SIDE ON 39, 66
errors in logical typing (see glossary) 3, 7, 49, 65
EVEN/ODD number of
– centres 8, 49, 51
– double bonds 5, 147, 203
– electron pairs 6, 8, 49
– methylene groups 30, 31, 34, 50, 52, 74, 77, 94
electrons, valence 16, 29, 32, 33, 44
evolution (see glossary)
experimental methods 157

Fe/Ni compared 58, 61
forbidden (see glossary)

GRIGNARD reactions 56, 76
GRIMM'scher Hybridverschiebungssatz 29, 143, 165

hand/gloves (see glossary)
hierarchical order 30, 81, 155, 156, 179
hyperplanes of 14
– concentrations 129
– properties 12

inhibited (see glossary)
intermodal (see glossary)
intermolecular (see glossary)
intramodal (see glossary)
intramolecular (see glossary)
inversion of local symmetry 178
IONIC/COVALENT 63, 65
ionization potentials, first 38
isoelectronic couplers 30, 31

keto/enole 8, 162
key/lock (see glossary)
KLOPMAN-SALEM theorem 65
KNOEVENAGEL addition 6, 8, 39, 59, 87–93
KOLBE-SCHMITT synthesis 43, 44

lacton rule 69
left (see glossary)
LINEAR/TILTED 39
local parity 54, 153, 154, 176

matched/mismatched pairs (see glossary) 103

membrane differentiations 43, 170, 171, 207
MENDEL's second law 174
metala-logy principle 54, 55, 57, 147
method of continuous variation 126, 134
MICHAEL addition 2, 6, 8, 39, 58, 74, 87–93
model calculations 112, 115
MUCH/A LITTLE 2, 75, 129, 131

Na/K compared 2, 86, 88, 89
negentropy (see glossary)
Ni/Pd compared 46, 106
Ni/Ti compared 2, 48
nitration 2, 5, 25, 30, 31, 34, 127
norcaradiene equilibrium 61
N/P compared 75, 129, 130

OCAMS method 148
O/C alkylation 50
oligomerization of
– aldehyde 106, 110
– diene 2, 12, 20, 46, 48, 59, 75, 114, 118, 129, 130, 131
– propene 121, 133
O/N compared 25, 30, 31, 36, 89
optical induction 5, 19, 20, 31, 61, 74, 92–103
ortho-/para-substitution 2, 5, 25, 30, 34, 44, 74
O/S compared 2, 17–19, 21, 22, 27, 39, 74, 84, 105, 106, 158, 159

parameters (see glossary)
parity (see glossary)
patterns (see glossary) 9, 15–17, 25, 27, 34, 45, 76, 77, 84, 88–90, 102, 105, 107, 154, 174
pattern comparison 25, 34, 174
PETERSON synthesis 63
PES data 21, 33
phase relation rules 148
phenyl/benzyl compared 2, 31, 34, 99
phenylogy principle 27, 55, 146
photocyclisation 50
^{31}P-NMR data 125, 126
poisons for catalysts 131
polymerization 40
position (see alternative position)
properties of
– ligands 9, 10–13, 20, 27, 36, 50, 58, 94, 101, 102, 106, 112
– solvents 54, 56, 84, 158
proteins 168
PSE-chess board 44
PSE sector rules 41

regioselectivity 2, 8, 39
representative substituents 18, 21, 22, 27, 181
repulsive (see glossary)
right (see glossary)
ring-closure by
– acids 6
– heating 6, 147
– metal-catalysts 55, 147
RNS 167
ROBINSON anellation 5, 6
R/S decisions in products 31, 97, 99–102

S/AS decisions 18, 52
selectivity def. 137
sectors in PSE 41
six-membered ring synthesis 6
specificity def. 137
sp^2/sp^3 hybridization 43
stereodifferentiation (see glossary)
stereoselectivity 2, 5, 6, 8
steric (see glossary)
"steric/electronic" 10, 65
"steric and electronic box" 12
strategies 157, 180
sympathetic pendulum 71
system enlargement 139, 142, 144, 145
system theory 137

tin compounds 79, 80
titration curves (see concentration control map) 111–113, 115

unifying principle (see direct unifying)
unified bonding 177

vinylogy principle 55, 146

WITTIG reaction 5, 27, 63, 74, 75, 78, 123–126

X-ray structures 36, 39, 40, 105

Zn/Cd compared 2, 158